Luca Ciotti
Osservatorio Astronomico di Bologna
Via Ranzani, 1
I-40127 Bologna
Italy
ciotti@bo.astro.it

Lecture Notes on Stellar Dynamics

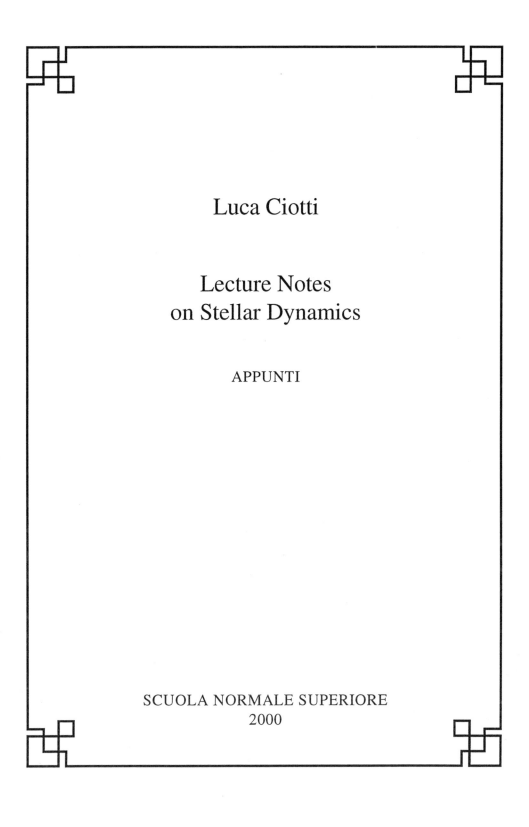

Luca Ciotti

Lecture Notes
on Stellar Dynamics

APPUNTI

SCUOLA NORMALE SUPERIORE
2000

FOREWORD

These Lecture Notes contain the main topics presented in a series of lectures given at the Scuola Normale di Pisa in the academic years 1998-1999-2000. The audience was made of students of the third and fourth year of the Physics program. In addition, the course was attended by a few graduate students of the Perfezionamento in Fisica. The lectures address selected topics in Stellar Dynamics, presented in a deductive approach.

Completeness is by no means the final goal of this work. The course is divided into two parts. In the first part (Chapters 1-8), basic mathematical concepts are introduced, with some discussion of their physical meaning. In the second part (Chapters 9-12), the tools developed in the first part are applied to an extensive discussion of selected problems offered by the Stellar Dynamics of collisionless systems of stars.

More specifically, in the first part, after a brief introduction dealing with the relations between Stellar Dynamics and the N-body problem, the main properties of ordinary differential equations are summarized. In Chapter 2, the Lagrangian derivative is introduced and the Transport theorem is derived in detail. In Chapter 3, some exact results on the N-body problem are derived and discussed, together with the main properties of singular solutions; furthermore, the behavior of special solutions of the N-body problem and the asymptotic behavior of general solutions for $t \to \infty$ are briefly considered. In Chapter 4, the concepts of microstate, macrostate, and ensemble are introduced, and the Liouville equation is derived. Using the method of characteristics it is then shown that the solution of the Liouville equation is equivalent to the solution of the original N-body problem. In Chapter 5 a detailed derivation of the 2-body relaxation time is given and two instructive cases are worked out explicitly. In Chapter 6 the concept of collisionless stellar system is clarified, and the Collisionless Boltzmann equation is derived, followed by a short discussion of the Fokker–Planck equation is also given. In Chapter 7 the method of Moments in velocity space is formulated in complete generality, from which the Jeans equations are deduced as special cases. The method of Moments is then applied to the Jeans equations in configuration space, leading to the Lagrange–Jacobi identity in tensorial form, for the general case where an external potential is also present. Finally, in Chapter 8, in view of applications to interpretation of observed data, the projection operator is introduced, together with the definition of the relevant velocity profiles for general systems. The relations among three basic sets of projected velocity moments are obtained, and the Projected Virial theorem is then formally derived.

The second part of these lectures is centered on three important problems of Stellar Dynamics, namely, on the concept of integrals of the motion, and on the two main approaches to the construction of stellar dynamical models, i.e., the "f–to–ρ" and the "ρ–to–f" methods (here called the direct and inverse problem of Stellar Dynamics, respectively). More specifically, in Chapter 9, a discussion of the concepts of local vs. global, regular and isolating integrals of the motion leads to the proof of the Jeans Theorem for stationary stellar systems. In Chapter 10, the direct problem of Stellar Dynamics is stated rigorously, and some applications to spherical and axisymmetric systems are presented. In Chapter 11, after a brief introduction to the Abel inversion technique for a special class of integral equations, the case of spherical and axisymmetric stellar systems is addressed, especially in view of the closure and integration of the associated Jeans equations. In Chapter 12, the inverse problem of Stellar Dynamics for spherical and axisymmetric systems is formulated and discussed in detail. A special technique to check the positivity of the supporting distribution function for spherically symmetric multi–component models, without resorting to an explicit derivation or numerical integration, is developed and demonstrated. These results are then applied to a family of spherically symmetric two–component anisotropic stellar systems.

In these Notes, vectors are indicated with bold–face letters, the symbol "$:=$" means "is defined as", while "$=$" is used for equality between two quantities defined previously; a dot over a symbol means a total time derivative; "$\sim$" means "asymptotic to", while "$\simeq$" indicates approximate numerical equivalence. Proofs of Theorems and Lemmas are in small text, so that they can be easily skipped by a reader interested only in the exposition. Finally, references are indicated by numbers in brackets.

Pisa, October 3, 2000   Luca Ciotti

ACKNOWLEDGEMENTS

I am very grateful to Giuseppe Bertin for his invitation to the Scuola Normale Superiore of Pisa and for enlightening discussions and useful advice on many aspects of this book; without his continuous encouragement this work could not have been completed. I am also grateful to many of the students who have taken this course, for their stimulating questions and comments. I would also like to thank Dr. Roberto Casini, Dr. Paolo Ciliegi and Dr. Alessandro Profeti for a very friendly help on TeX, in which these Notes have been written. I am especially grateful to Giuseppe Bertin, James Binney, Jerry Ostriker, and Alvio Renzini from which I learned most of what I know in Astrophysics.

Finally, I would like to thank my wife and colleague, Silvia Pellegrini, because without her love and patience this book would not exist: these Lectures are dedicated to her and to our little daughter Anna Magdalena.

CONTENTS

1. STELLAR DYNAMICS

We will follow a constructive approach towards the derivation of many fundamental results of the subject known as Stellar Dynamics. This introductory Chapter provides a simple and qualitative discussion of the general goals of Stellar Dynamics. We focus on the astrophysical plausibility of the model that we pose at the basis of our treatment, the so–called N-body problem. This model considers a system of N point masses in mutual gravitational interaction as the starting point to understand the evolution of large stellar systems, such as globular clusters and galaxies. In these Notes we will therefore neglect all the phenomena related to other factors, in particular to the presence of interstellar gas and electromagnetic fields.

1.1 What is Stellar Dynamics ?

Stellar Dynamics is a broad field of study, with no simple definition. Hopefully, an idea of a (small) selection of problems and methods encountered in this field can be gained by following these Lecture Notes. These form by no means a complete review. They are meant to be used as a concise but rigorous introduction to the subject for the interested reader: a more in-depth treatment of several of the themes presented here can be found, for example, in [5.1], [5.2], [5.4], [6.1], [6.3], [6.9].

We start by stating that the main tasks of Stellar Dynamics are to obtain a *qualitative understanding of the* **structure** *and the* **evolution** *of stellar systems (e.g., open and globular clusters, galaxies, clusters of galaxies) and to develop* **mathematical methods** *(analytical and numerical) capable of quantitative predictions.*

Stellar Dynamics is not a self–contained research field, because it makes use of techniques and results borrowed from other fields of Physics and Mathematics. As a way to set a theme, we may state that Stellar Dynamics deals with gravitational systems made of a large number of "particles" (N-body systems, with, as a reference number, $N > 10$). The connections of Stellar Dynamics with Celestial Mechanics ($N \leq 10$) are – as expected – very strong, but also Analytical Mechanics, Statistical Mechanics, Fluid Dynamics, and Plasma Physics share techniques and results with Stellar Dynamics.

Obviously, we should not forget the importance of observations, which are the true contact point between theory and reality. Observations play a two–fold role in Stellar Dynamics. On the one hand, the theory can serve as a source of interesting objectives for the observers, challenging their skills and instruments; on the other hand, observations of increasing accuracy can confirm or disprove theoretical results, sometimes presented initially without adequate empirical support.

Stellar dynamicists should recognize *explicitly* the special nature of the assumptions made and continuously revise the relations between the adopted model and the actual astrophysical system that is addressed.

1.2 Stellar Dynamics and the N-body problem

One of the most important problems of Stellar Dynamics concerns the formal treatment of the building blocks of any stellar system, i.e., the stars.

Let us consider the solar radius $R_\odot$ ($R_\odot \simeq 7 \times 10^{10}$ cm) as the characteristic radius of a star in a stellar system, and, moreover, let us take a number

N of stars *homogeneously* distributed inside a sphere of radius R. The *geometrical* cross–section for impacts between two stars is obviously $\sigma_* = 4\pi R_\odot^2$; thus a star is associated with a volume given by $\lambda_g \sigma_*$, where, by definition,

$$N\lambda_g\sigma_* = \frac{4\pi}{3}R^3, \qquad (1.1)$$

i.e.,

$$\frac{\lambda_g}{2R} = \left(\frac{R}{R_\odot}\right)^2 \frac{1}{6N}. \qquad (1.2)$$

The quantity λ_g provides an order–of–magnitude estimate of the path that a star can traverse without suffering geometrical collisions with another star of the system. We now assume as characteristic values for N and R the pairs $(10^{11}; 10^4 \text{ pc})$ and $(10^6; 10 \text{ pc})$. Here we recall that $1 \text{ pc} = 3.08 \times 10^{18}$ cm. These may be taken as representative values for an elliptical galaxy and a globular cluster, respectively. In the first case $\lambda_g/2R \simeq 3 \times 10^{11}$, and, in the second, $\lambda_g/2R \simeq 3 \times 10^{10}$. Astronomers tell us that the characteristic velocities of stars (v_*) in elliptical galaxies and in globular clusters are of the order of few hundreds km s^{-1}, and of few tens km s^{-1}, respectively. Even if we assume that elliptical galaxies and globular clusters are as old as the Universe $(t_H \simeq 15 \text{ Gyr} \simeq 4.7 \times 10^{17} \text{ s})$, we obtain $\lambda_g/v_* t_H \approx 10^9$ for elliptical galaxies and $\lambda_g/v_* t_H \approx 10^6$ for globular clusters. *As a consequence, in stellar systems such as elliptical galaxies and globular clusters, one can reasonably exclude the occurrence of geometrical collisions between stars over time–scales much longer than the present age of the Universe.* It is important to note that in some astrophysical situations, e.g., during the so–called *core–collapse* of globular clusters, or in regions of very high stellar density as in galactic nuclei, the probability of physical collisions between stars may be not so small, and this type of direct stellar collisions can play some role in the evolution of those systems.

This qualitative discussion justifies the assumption made in the majority of problems investigated in Stellar Dynamics, i.e., that geometrical collisions between stars are ignored. Each star is then assumed to be a *spherically symmetric* point mass so that the starting point of Stellar Dynamics is represented by the differential equations describing probably the most famous problem of all Mathematical Physics, i.e., the N-body problem. Here we briefly recall how this problem is formulated, both in its Newtonian and in its Hamiltonian form.

In the Newtonian formulation (in Cartesian coordinates), the equations of motion for each of the N particles, expressed in an inertial frame of reference S_0, are given by:

$$\begin{cases} \dot{\mathbf{x}}_i = \mathbf{v}_i, \\ \dot{\mathbf{v}}_i = -\dfrac{1}{m_i}\dfrac{\partial U}{\partial \mathbf{x}_i}, \\ U = -\dfrac{G}{2}\displaystyle\sum_{i\neq j=1}^{N}\dfrac{m_i m_j}{\|\mathbf{x}_i - \mathbf{x}_j\|}, \end{cases} \qquad (1.3a)$$

for $i = 1,\ldots,N$, where m_i, $\mathbf{x}_i$, and $\mathbf{v}_i$ are mass, position, and velocity of the i-th star, U is the potential energy, $\partial U/\partial \mathbf{x}_i$ is the gradient of U with respect to $\mathbf{x}_i$, and finally $\|\mathbf{x}\| := \sqrt{<\mathbf{x},\mathbf{x}>} = \sqrt{x_1^2 + x_2^2 + x_3^2}$ is the standard Euclidean norm. The constant $G = 6.67 \times 10^{-8}$ cm^3 s^{-2} g^{-1} is the gravitational constant. The problem is usually formulated with the assignment of the *initial conditions*:

$$\begin{cases} \mathbf{x}_i(0) = \mathbf{x}_i^0, \\ \mathbf{v}_i(0) = \mathbf{v}_i^0, \end{cases} \qquad (1.3b)$$

for $i = 1,\ldots,N$.

In the Hamiltonian formulation, the N-body problem is given by

$$\begin{cases} \dot{\mathbf{q}}_i = \dfrac{\partial H}{\partial \mathbf{p}_i}, \\ \dot{\mathbf{p}}_i = -\dfrac{\partial H}{\partial \mathbf{q}_i}, \\ H = T + U = \displaystyle\sum_{i=1}^{N}\dfrac{\|\mathbf{p}_i\|^2}{2m_i} - \dfrac{G}{2}\sum_{i\neq j=1}^{N}\dfrac{m_i m_j}{\|\mathbf{q}_i - \mathbf{q}_j\|}, \end{cases} \qquad (1.4)$$

where $(\mathbf{q}_i, \mathbf{p}_i) := (\mathbf{x}_i, m_i\mathbf{v}_i)$ for $i = 1,\ldots,N$, T is the total kinetic energy; the initial conditions are given by $2N$ relations equivalent to eq. (1.3b).

As is well known, the properties of the general solution of the N-body problem are not presently available, which makes this a problem of great mathematical interest. The various techniques developed in order to extract information from the previous set of differential equations have contributed in an essential way to the achievement of many fundamental results, both in pure and applied mathematics, also in the context of numerical analysis. Given the enormous amount of work done in the last three centuries on this subject, it is impossible to provide here even a partial review of the most important results. Nevertheless, given their importance in Stellar Dynamics, in the following Chapters I will try to summarize at least some definitions and some basic theorems characterizing the key issues in the N-body problem.

2. BASIC METHODS

This Chapter describes in detail the basic mathematical methods used in Stellar Dynamics. The focus is on the Transport theorem, which lie at the foundation of the most important equations used as tools in the study of the dynamics of stellar systems. Many results will follow from the use of specific mathematical techniques introduced here. Geometrical arguments often help the intuition towards results that are proved separately by rigorous analytical derivations.

2.1 Fundamental properties of ODEs

We start by introducing the standard definition of Ordinary Differential equations in $\Re^n$:

Definition 2.1 [ODE] *Let $A \subseteq \Re^n$ be an open set and $\mathbf{W} : A \mapsto \Re^n$ a vector field. The system*

$$\begin{cases} \dot{\mathbf{x}} := \dfrac{d\mathbf{x}}{dt} = \mathbf{W}(\mathbf{x}) \\ \mathbf{x}(0) = \mathbf{x}^0, \end{cases} \qquad (2.1a)$$

is called Ordinary Differential equation *(ODE) in $\Re^n$ with initial condition $\mathbf{x}^0$. An application $\mathbf{\Psi} : A \times I \mapsto \Re^n$, where $I =]-\delta, \delta[$ and $\delta > 0$, so that*

$$\begin{cases} \dot{\mathbf{\Psi}}(\mathbf{x}^0; t) = \mathbf{W}[\mathbf{\Psi}(\mathbf{x}^0; t)], \\ \mathbf{\Psi}(\mathbf{x}^0; 0) = \mathbf{x}^0, \end{cases} \qquad (2.1b)$$

is called the solution with initial condition $\mathbf{x}^0$ for the ODE given in eq. (2.1a). The space $\Re^n$ is called phase–space, *and $\Re^n \times \Re$ the* extended phase–space. *Usually δ is finite: otherwise the solution is said to be a* flow.

In Mathematical Physics t is usually regarded as the *time*; from the previous definition, the coordinates of the system in phase–space at an arbitrary time t are given by the *time section* of the extended phase–space at that time.

It is important to note that the previous definition of ODE describes only the *autonomous* ODEs, i.e., the ODEs for which the associated vector field $\mathbf{W}$ is independent of time. Even *non–autonomous* ODEs [for which $\mathbf{W} = \mathbf{W}(\mathbf{x}; t)$] can be included in Definition 2.1, provided we increase the dimension of phase–space. This is done by introducing a fictitious time τ, related to t by the differential equation $dt/d\tau = 1$, by interpreting t as the $(n+1)$–th component of the extended vector $(\mathbf{x}, t)$, and finally by extending the vector field $\mathbf{W}$ [an $(n+1)$–th component equal to 1 is added].·

The central problem of the theory of ODEs is to determine whether a solution exists and what are its properties. Many theorems are available for most cases of interest (see, e.g., [1.3], [2.5], [4.1]). Here we report only some basic results.

Theorem 2.2 [Local existence, uniqueness, and regularity of solutions]
*Following the notation introduced in Definition 2.1, let $U_0 \subseteq A$, with $\mathbf{x}^0 \in U_0$,
be an open set. $C^0(U_0)$ is the set of functions continuous on U_0; $C^r(U_0)$ is
the set of functions with the r–th derivative continuous on U_0.*
*I) If $\mathbf{W} \in C^0(U_0)$, then the solution exists at least locally, i.e., one can find
a $\delta > 0$ for which eq. (2.1b) holds[2.1].*
IIa) If $\mathbf{W}$ is Lipschitz on U_0, i.e., if there exists $K_0 > 0$ so that

$$\|\mathbf{W}(\mathbf{x}_1^0) - \mathbf{W}(\mathbf{x}_2^0)\| \leq K_0\|\mathbf{x}_1^0 - \mathbf{x}_2^0\|, \quad \forall\, (\mathbf{x}_1^0, \mathbf{x}_2^0) \in U_0, \tag{2.2a}$$

then $\delta > 0$ exists so that for $t \in\,] - \delta, \delta[$ the solution (exists and) is unique.
*IIb) In addition, under the assumptions of point IIa), if $\boldsymbol{\Psi}(\mathbf{x}_1^0; t)$ and $\boldsymbol{\Psi}(\mathbf{x}_2^0; t)$
are solutions of eq. (2.1a), then $\forall t \in I =\,] - \delta, \delta[$*

$$\|\boldsymbol{\Psi}(\mathbf{x}_1^0; t) - \boldsymbol{\Psi}(\mathbf{x}_2^0; t)\| \leq \|\mathbf{x}_1^0 - \mathbf{x}_2^0\|e^{K_0\, t}, \tag{2.2b}$$

i.e., the solution depends continuously on the initial conditions.
*III) If $\mathbf{W} \in C^r(U_0)$, with $r \geq 1$, then points I), II) above are satisfied,
$\boldsymbol{\Psi}(\mathbf{x}^0, t) \in C^r(U_0)$ with respect to $\mathbf{x}^0$, and $\boldsymbol{\Psi}(\mathbf{x}^0, t) \in C^{r+1}(I)$ with respect to
t.*

Note that point II) has an important consequence: where the solution exists
and is unique, necessarily the solution itself is *time–invertible* (at least in
principle), i.e., $\forall t \in\,] - \delta, \delta]$ and $\forall \mathbf{x}^0 \in U_0$,

$$\begin{cases} \mathbf{x} = \boldsymbol{\Psi}(\mathbf{x}^0; t), \\ \mathbf{x}^0 = \boldsymbol{\Psi}[\boldsymbol{\Psi}(\mathbf{x}^0; t); -t]. \end{cases} \tag{2.3}$$

Therefore[2.2], with a suggestive notation we can write $\boldsymbol{\Psi}^{-1}(\mathbf{x}; t) := \boldsymbol{\Psi}(\mathbf{x}; -t)$.
From a physical point of view, this means that it is possible to integrate
the ODE backward in time. Moreover, if condition III) applies, by defining
J_0 the determinant of the Jacobian matrix of the coordinate transformation
$\mathbf{x} = \boldsymbol{\Psi}(\mathbf{x}^0; t)$ with respect to $\mathbf{x}^0$,

$$J_0 := \det\left[\frac{\partial \boldsymbol{\Psi}(\mathbf{x}^0; t)}{\partial \mathbf{x}^0}\right], \tag{2.4a}$$

[2.1] How large δ can be depends on the specific behaviour of $\mathbf{W}$ near $\mathbf{x}^0$ (see,
e.g., [2.5]).
[2.2] The result in eq. (2.3) is easily proved for a flow, due to the fact that
in this case Ψ is a so–called 1–parameter (t) continuous group, for which
$\forall (t_1, t_2)\ \boldsymbol{\Psi}[\boldsymbol{\Psi}(\mathbf{x}; t_1), t_2] = \boldsymbol{\Psi}(\mathbf{x}, t_1 + t_2)$. See, e.g., [4.1].

$$\left[\frac{\partial \boldsymbol{\Psi}(\mathbf{x}^0;t)}{\partial \mathbf{x}^0}\right]_{ij} := \frac{\partial \Psi_i(\mathbf{x}^0;t)}{\partial x_j^0}; \quad (i,j=1,\ldots,n), \tag{2.4b}$$

one immediately derives that $J_0 \neq 0$ over $U_0 \times I$. Finally, from the fact that $J_0(0) = 1$, it follows that $J_0(t)$ remains positive for $t \in I$.

2.2 Material derivative

A key concept associated with the solution of the ODE given in Definition 2.1 is identified by considering the behaviour of a given scalar function $f : \Re^n \times \Re \mapsto \Re$, $f = f(\mathbf{x};t)$. For each initial condition $\mathbf{x}^0$ – through the solution $\boldsymbol{\Psi}(\mathbf{x}^0;t)$ – it is possible to introduce a new function $f_{\mathcal{L}}(\mathbf{x}^0;t) := f[\boldsymbol{\Psi}(\mathbf{x}^0;t);t]$. From a physical point of view, this function describes how the "property" f – which, at time $t = 0$, is associated with the initial condition $\mathbf{x}^0$ – evolves with time. The time evolution of $f_{\mathcal{L}}$ is obtained by introducing the following definition (see, e.g., [3.1], [3.2], [3.3]):

Definition 2.3 [Material derivative in Lagrangian form] *The material derivative in Lagrangian form (or, simply, Lagrangian derivative) $\mathcal{L}_{\mathbf{W}}(f)$ of f along the vector field $\mathbf{W}$ with initial condition $\mathbf{x}^0$ is defined as*

$$\mathcal{L}_{\mathbf{W}}(f) := \frac{df_{\mathcal{L}}(\mathbf{x}^0;t)}{dt}, \tag{2.5}$$

where d/dt is the standard derivative.

From this definition, it could be erroneously believed that – for practical purposes – the material derivative is of no use, for it presumes the knowledge of the solution of the associated ODE for a generic initial condition. This is only partially true. In fact:

Theorem 2.4 [Material derivative in Eulerian form] *If $f \in C^r(\Re^n)$ ($r \geq 1$), the following identity holds:*

$$\mathcal{L}_{\mathbf{W}}(f) = \left(\frac{Df}{Dt}\right)_{\mathbf{x}=\boldsymbol{\Psi}(\mathbf{x}^0;t)}, \tag{2.6a}$$

with

$$\frac{Df}{Dt} := \frac{\partial f}{\partial t} + \left< \mathbf{W}, \frac{\partial f}{\partial \mathbf{x}} \right>. \tag{2.6b}$$

As a consequence, the *functional* form of the material derivative is independent of the choice of $\mathbf{\Psi}$, but depends only on $\mathbf{W}$. The differential operator D/Dt is sometimes called "total" derivative, even though the correct (but perhaps less suggestive) definition is that of *material derivative in Eulerian form*. Two examples of explicit Lagrangian derivatives useful in applications are now presented.

Let us consider the motion of a particle in a conservative force field in $\Re^3$ with potential ϕ, $\ddot{\mathbf{x}} = -\partial\phi/\partial\mathbf{x}$. The phase–space for this system is $\Re^6$, and, according to Definition 2.1, $\mathbf{W} = (\mathbf{v}, -\partial\phi/\partial\mathbf{x})$. From eq. (2.6b)

$$\frac{Df}{Dt} = \frac{\partial f}{\partial t} + <\mathbf{v}, \frac{\partial f}{\partial\mathbf{x}}> - <\frac{\partial\phi}{\partial\mathbf{x}}, \frac{\partial f}{\partial\mathbf{v}}> . \tag{2.7}$$

The example above is a special case of the more general mechanical system described by the Hamiltonian function $H = H(\mathbf{q}, \mathbf{p})$. In this case we have $\mathbf{W} = (\partial H/\partial\mathbf{p}, -\partial H/\partial\mathbf{q})$, and so

$$\frac{Df}{Dt} = \frac{\partial f}{\partial t} + <\frac{\partial H}{\partial\mathbf{p}}, \frac{\partial f}{\partial\mathbf{q}}> - <\frac{\partial H}{\partial\mathbf{q}}, \frac{\partial f}{\partial\mathbf{p}}> = \frac{\partial f}{\partial t} + [f, H], \tag{2.8}$$

where $[f, H]$ is the *Poisson bracket* between f and H (see Definition 9.6).

As will become clear in the following developments, the importance of a Eulerian version of the material derivative (i.e., an expression in terms of the independent coordinates of phase–space) is of great importance. In fact, especially when dealing with continuous systems, while the Lagrangian description readily leads to a simple set of evolutionary equations, the Eulerian formulation is best suited for direct applications.

In the next Section the results obtained so far for a *specific* initial condition $\mathbf{x}^0$ will be extended to a whole collection of initial conditions.

2.3 The Transport theorem

We now prove a very general result, which is the basis for a unified derivation of the differential equations describing the motion of *continuous systems*, such as, for example, those of Fluid Dynamics, the motion of a set of points in phase–space under the action of phase flows, the evolution of the so–called *macroscopic properties* of stellar systems.

Let us consider an arbitrary region $A(0)$ of phase–space at $t = 0$. The phase–flow $\mathbf{\Psi}(\mathbf{x}^0; t)$ induced by the vector field $\mathbf{W}$ "moves" the set $A(0)$ in the extended phase–space, transforming $A(0)$ into $A(t)$. The function

$$F(t) := \int_{A(t)} f(\mathbf{x}; t) d^n\mathbf{x} \tag{2.9a}$$

is the integral of f over the region $A(t)$ (see Fig. 2.1). We look for the time variation of $F(t)$, i.e., we wish to determine the explicit expression of dF/dt. The standard method is based on the assumption of uniqueness of the solutions associated with the vector field $\mathbf{W}$, and, as a consequence, on the possibility to change variables $\mathbf{x} \mapsto \mathbf{x}^0$ in the integral representation for $F(t)$, as described by eq. (2.3).

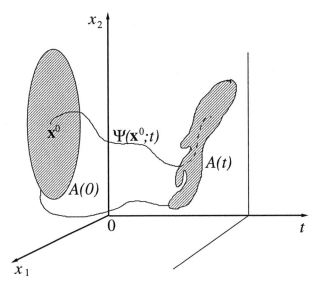

Figure 2.1

Therefore:

$$F(t) = \int_{A(0)} f_{\mathcal{L}}(\mathbf{x}^0; t) J_0(t) d^n \mathbf{x}^0, \qquad (2.9b)$$

where J_0 is given by eqs. (2.4ab). Note that in eq. (2.9) J_0 appears instead of $|J_0|$, because J_0 is positive definite. Before we embark in the derivation of the theorem, we prove the following

Lemma 2.5 *For J_0 defined by eqs. (2.4ab) the following identity holds $\forall t \in$ $]-\delta, \delta[$:*

$$\frac{dJ_0}{dt} = J_0 \times \operatorname{div}_{\mathbf{x}} \mathbf{W}\big|_{\mathbf{x}=\boldsymbol{\Psi}(\mathbf{x}^0; t)}. \qquad (2.10)$$

PROOF　　The proof of eq. (2.10) is based on the well–known property of determinants under derivation with respect to a parameter (in our case the time t). In fact, the derivative of a determinant of an $n \times n$ matrix is found to be equal to the sum of n determinants, where in each determinant a different row (column) of the original matrix is replaced by its derivative with respect to the parameter (see, e.g., [1.4], [3.1], [3.2], [3.3]).

Let us compute explicitly the first of the n determinants associated with the time differentiation of J_0, a determinant where the first row has been replaced by its derivative, and all other rows are unchanged with respect to the original matrix given by eq. (2.4b):

$$\frac{d}{dt}\frac{\partial\Psi_1}{\partial x_j^0} = \frac{\partial}{\partial x_j^0}\frac{\partial\Psi_1}{\partial t} = \frac{\partial W_1^0}{\partial x_j^0},$$

where $W_1^0 := W_1[\boldsymbol{\Psi}(\mathbf{x}^0;t)]$. The last identity follows from eq. (2.1b) directly. As a result, the j-th element in the first row is:

$$\frac{\partial W_1^0}{\partial x_j^0} = \frac{\partial W_1}{\partial\Psi_1}\frac{\partial\Psi_1}{\partial x_j^0} + \sum_{i=2}^{n}\frac{\partial W_1}{\partial\Psi_i}\frac{\partial\Psi_i}{\partial x_j^0}.$$

The sum on the r.h.s. of this equation is just a linear combination of the remaining $i = 2,\ldots,n$ elements of column j, each multiplied by the coefficient $\partial W_1/\partial\Psi_i$. Therefore, the first row of the determinant is given by the vector $(\partial W_1/\partial\Psi_1)(\partial\Psi_1/\partial x_j^0)$ plus a linear combination of the remaining rows, each weighted by $\partial W_1/\partial\Psi_i$. The determinant is independent of such linear combination, and it amounts to $J_0\partial W_1/\partial\Psi_1$. The same argument applies to the remaining $n-1$ determinants, and so, after summation, eq. (2.10) is proved. ◁

Theorem 2.6 [Transport theorem] *Under the conditions and notation described so far,*

$$\frac{dF(t)}{dt} = \int_{A(t)}\left[\frac{Df}{Dt} + f\,\mathrm{div}_{\mathbf{x}}\mathbf{W}\right]d^n\mathbf{x}. \tag{2.11}$$

PROOF   Differentiating eq. (2.9b) with respect to time, and using Lemma 2.5 one obtains

$$\frac{dF(t)}{dt} = \int_{A(0)}\left[\frac{df_{\mathcal{L}}}{dt} + f_{\mathcal{L}}\mathrm{div}_{\mathbf{x}}\mathbf{W}\big|_{\mathbf{x}=\boldsymbol{\Psi}(\mathbf{x}^0;t)}\right]J_0(t)d^n\mathbf{x}^0.$$

By going back to the original integration variables $\mathbf{x}$, the theorem is proved.   ◁

An important application of the previous result is obtained under the assumption that f is a *conserved* quantity, i.e., $dF(t)/dt = 0$ for all choices of the control volume $A(0)$. In this case, we have immediately

Theorem 2.7 *If $dF(t)/dt = 0$ for any $A(0)$, then*

$$\frac{Df}{Dt} + f\,\mathrm{div}_{\mathbf{x}}\mathbf{W} = \frac{\partial f}{\partial t} + \mathrm{div}_{\mathbf{x}}(f\mathbf{W}) = 0; \tag{2.12a}$$

moreover, if $\mathbf{W}$ is solenoidal, i.e., if $\mathrm{div}_{\mathbf{x}}\mathbf{W} = 0$ over the phase–space, then

$$\frac{Df}{Dt} = 0. \tag{2.12b}$$

It is important to note that the conservation of f inside a prescribed volume means that no sources or sinks are present. In fact, for the uniqueness of the solution of a well–behaved ODE, it is impossible for a point inside the volume to "escape" outside. Otherwise at some time the trajectory of a phase–space point would have to cross the boundary of the volume, violating the uniqueness of the solution[2.3]. As a simple application of the previous results, we derive the

Theorem 2.8 [Liouville theorem] *The volume in phase–space of a Hamiltonian system is conserved.*

PROOF   With the choice $f = 1$ in eq. (2.9a), $F(t) = A(t)$. Given the fact that the Hamiltonian flow is solenoidal, the Liouville theorem is a direct consequence of eq. (2.11). For a discussion of this fundamental result in various contexts see, e.g., [1.1], [1.3], [1.5], [5.3], [5.6], [5.7].   ◁

[2.3] At the time of contact, at a point, between a trajectory and the boundary, the solution would not be unique.

3. EXACT RESULTS ON THE N-BODY PROBLEM

In this Chapter, after a short review of the main properties of N-body systems, we focus on the exact solutions of the N-body problem. Such discussion is not found frequently in textbooks on Stellar Dynamics. Within the very large set of results available on the N-body problem, many are relatively easy to derive and to understand and, because of this, should deserve wider attention. Some of these are presented in this Chapter, because the N-body problem is at the basis of Stellar Dynamics and because we wish to point out that some results commonly used can be derived directly from the equations of the motion for the N-body problem, without any need for the more sophisticated mathematical techniques that will be developed in the next chapters.

3.1 A short review

Before undertaking the classical approach to Stellar Dynamics, three important aspects of the *exact* treatment of the N-body problem should be recalled. Here "exact" means *derived from the differential equations of the N-body problem together with the appropriate initial conditions.* As we will see, this does not mean that other approaches are *incorrect*, but, rather, that in alternative approaches characteristic of Stellar Dynamics, some *additional assumptions* are explicitly (or implicitly) made. Our task will be to state the necessary additional assumptions explicitly. Moreover, this short presentation should convince the reader of the *need* for alternative approaches (instead of a direct attack to the N-body differential equations) to the problems addressed by Stellar Dynamics.

As is well known, for $N \geq 3$ the N-body problem is generally not *solvable* in the classical meaning of the word, i.e., *it cannot be reduced to $6N - 1$ independent integrations* (see, e.g., [1.7]). Nevertheless, many important properties of N-body systems can be explored analytically by a direct use of eqs. (1.3ab).

Let us start by introducing some standard notation. In the following, S_0 is an inertial frame of reference.

Definition 3.1 *For a generic N-body system, we define:*

$$r_{ij} := ||\mathbf{x}_i - \mathbf{x}_j||, \quad r := \min_{i \neq j} r_{ij}, \quad R := \max_{i,j} r_{ij}, \qquad (3.1a)$$

$$M := \sum_{i=1}^{N} m_i, \quad \mu := \min_i m_i, \qquad (3.1b)$$

$$V := -U, \quad T := \frac{1}{2}\sum_{i=1}^{N} m_i ||\mathbf{v}_i||^2, \quad I := \sum_{i=1}^{N} m_i ||\mathbf{x}_i||^2, \qquad (3.1c)$$

$$M\mathbf{R}_{\mathrm{CM}} := \sum_{i=1}^{N} m_i \mathbf{x}_i, \quad M\mathbf{V}_{\mathrm{CM}} := \sum_{i=1}^{N} m_i \mathbf{v}_i. \qquad (3.1d)$$

M is the total mass, T and U (see eq. [1.3a]) are the kinetic and potential energy, respectively. I is the polar moment of inertia, $\mathbf{R}_{\mathrm{CM}}$ and $\mathbf{V}_{\mathrm{CM}}$ are the position and velocity of the center of mass, respectively. V is sometimes called the binding energy.

Theorem 3.2 [The 7 classical integrals[3.1]] *In the inertial frame of reference S_0 the following quantities are conserved:*

$$E := T + U, \quad \mathbf{P} := \sum_{i=1}^{N} m_i \mathbf{v}_i, \quad \mathbf{L} := \sum_{i=1}^{N} m_i \mathbf{x}_i \wedge \mathbf{v}_i, \tag{3.2}$$

where E, $\mathbf{P}$, and $\mathbf{L}$ are the total energy, momentum, and angular momentum, respectively.

PROOF   From eqs. (3.2) and (1.3a) it is easily shown that

$$\frac{dE}{dt} = \sum_{i=1}^{N} < \dot{\mathbf{x}}_i, m_i \ddot{\mathbf{x}}_i + \frac{\partial U}{\partial \mathbf{x}_i} >= 0,$$

$$\frac{d\mathbf{P}}{dt} = \sum_{i=1}^{N} m_i \ddot{\mathbf{x}}_i = -\frac{G}{2} \sum_{i,j=1;j\neq i}^{N} \frac{m_i m_j (\mathbf{x}_i - \mathbf{x}_j)}{r_{ij}^3} = 0,$$

$$\frac{d\mathbf{L}}{dt} = \sum_{i=1}^{N} m_i (\dot{\mathbf{x}}_i \wedge \dot{\mathbf{x}}_i + \mathbf{x}_i \wedge \ddot{\mathbf{x}}_i) = -\frac{G}{2} \sum_{i,j=1;j\neq i}^{N} \frac{m_i m_j \mathbf{x}_i \wedge (\mathbf{x}_i - \mathbf{x}_j)}{r_{ij}^3} = 0.$$

◁

A particularly important reference system associated with a generic N-body system is the (barycentric) inertial frame S', with origin located at $\mathbf{R}_{\mathrm{CM}}$ and axes parallel to the axes of S_0 at every time. By definition,

$$\mathbf{x}_i = \mathbf{x}'_i + \mathbf{R}_{\mathrm{CM}}, \quad \mathbf{v}_i = \mathbf{v}'_i + \mathbf{V}_{\mathrm{CM}}, \tag{3.3a}$$

and so in S_0

$$\mathbf{P} = M\mathbf{V}_{\mathrm{CM}}, \quad \mathbf{L} = M\mathbf{R}_{\mathrm{CM}} \wedge \mathbf{V}_{\mathrm{CM}} + \mathbf{L}', \tag{3.3b}$$

$$T = \frac{M\|\mathbf{V}_{\mathrm{CM}}\|^2}{2} + T', \quad U = U', \quad E = \frac{M\|\mathbf{V}_{\mathrm{CM}}\|^2}{2} + E', \tag{3.3c}$$

$$I = M\|\mathbf{R}_{\mathrm{CM}}\|^2 + I'. \tag{3.3d}$$

Note that $M\mathbf{R}_{\mathrm{CM}} \wedge \mathbf{V}_{\mathrm{CM}}$ is a time–independent vector. Due to the time–independence of the total angular momentum $\mathbf{L}$, the following geometrical concept is useful in applications:

[3.1] In several textbooks it is written that the number of classical integrals of the N-body problem is 10. For a discussion of this point see Chapter 9.

Definition 3.3 [Invariable plane] *If* $\mathbf{L}' \neq 0$, *the plane defined in* S' *as*

$$\Pi_{\text{inv}} := \{\mathbf{x}' \in \Re^3 : <\mathbf{x}', \mathbf{L}'>= 0\}, \tag{3.4}$$

is called the invariable plane.

The two following results find frequent applications:

Theorem 3.4 [Lagrange–Jacobi identity] *In* S_0

$$\ddot{I} = 2(2T + U) = 2(T + E) = 2(2E - U). \tag{3.5}$$

PROOF   The second and third identities derive from the first identity using energy conservation. The first identity is obtained by time differentiation of eq. (3.1c): $\ddot{I} = 2\sum_{i=1}^{N} m_i(<\dot{\mathbf{x}}_i, \dot{\mathbf{x}}_i> + <\mathbf{x}_i, \ddot{\mathbf{x}}_i>)$. As can be seen from eq. (1.3a), U is a homogenous function of order -1. Using the Euler theorem on homogeneous functions, the result is proved.   ◁

Lemma 3.5 *The polar momentum of inertia in the frame* S' *can be expressed as*

$$I' = \frac{1}{2M} \sum_{i,j=1}^{N} m_i m_j r_{ij}^2. \tag{3.6}$$

PROOF   From eq. (3.1d) the identity $\sum_{i=1}^{N} m_i r_{ij}^2 = I' + M||\mathbf{x}_j'||^2$ follows. After multiplication of both sides by m_j, and summation over j, the result is obtained.   ◁

Finally, two important estimates of the minimum and maximum separation between particles in a N-body problem can be obtained, as stated in the following

Theorem 3.6 *The following estimates hold* $\forall t$:

$$\frac{A}{V(t)} \leq r(t) \leq \frac{B}{V(t)}, \tag{3.7a}$$

$$C\sqrt{I'(t)} \leq R(t) \leq D\sqrt{I'(t)}, \tag{3.7b}$$

where A, B, C, D *are positive constants dependent only on the masses* m_i.
PROOF   The constants A and B are determined by using the binding energy V. In fact,

$$V = \frac{G}{2} \sum_{i,j=1;j\neq i}^{N} \frac{m_i m_j}{r_{ij}} \geq \frac{G m_i m_j}{r_{ij}}$$

$\forall i, j = 1, ..., N$ & $i \neq j$. In particular, the previous inequality holds for the pair of particles at the minimum relative distance, $(m_{i'}; m_{j'})$:

$$V \geq \frac{G m_{i'} m_{j'}}{r} \geq \frac{G\mu^2}{r},$$

and so $A = G\mu^2$. Moreover,

$$V = \frac{G}{2} \sum_{i,j=1; j \neq i}^{N} \frac{m_i m_j}{r_{ij}} \leq \frac{G}{2r} \sum_{i,j=1; j \neq i}^{N} m_i m_j \leq \frac{G}{2r} \sum_{i,j=1}^{N} m_i m_j = \frac{GM^2}{2r},$$

and so $B = GM^2/2$. The constants C and D are determined by using the polar moment of inertia in S' as given by eq. (3.6):

$$I' = \frac{1}{2M} \sum_{i,j=1}^{N} m_i m_j r_{ij}^2 \leq \frac{R^2}{2M} \sum_{i,j=1}^{N} m_i m_j = \frac{MR^2}{2},$$

and so $C = \sqrt{2/M}$. Moreover,

$$I' = \frac{1}{2M} \sum_{i,j=1}^{N} m_i m_j r_{ij}^2 \geq \frac{\mu^2}{2M} \sum_{i,j=1}^{N} r_{ij}^2 \geq \frac{\mu^2 R^2}{M},$$

and so $D = \sqrt{M}/\mu$.   ◁

3.2 Singularities

The investigation of the singularities of the ODEs describing the N-body problem (eq. [1.3a]) is an expansive field of research, started in 1895 by the French mathematician Paul Painlevé (1863–1933). The main problem posed by eq. (1.3a) is the fact that, since these equations are non–linear, their solutions can develop *movable* singularities, i.e., singularities that depend on the specific choice of initial conditions (eq. [1.3b]). In contrast, linear ODEs can present only *fixed* singularities, i.e., singularities that are independent of the initial conditions. In order to better understand the difference between the two cases, let us start with the following

Definition 3.7 [Homogeneous linear ODE] *An ODE of the form*

$$\frac{d^n y}{dx^n} + \sum_{i=0}^{n-1} p_i(x) \frac{d^i y}{dx^i} = 0, \tag{3.8}$$

where $d^0 y/dx^0 := y$ and $p_i(x)$ are functions of the independent variable, is called a homogeneous linear ODE of order n.

It is well known that all singularities exhibited by the solution of an ODE of the previous form are *only* associated with the singularities of the coefficients, and, therefore, they are independent of the initial conditions. This makes it possible to set up a complete classification of the singularities *without* solving the equation itself. The reader interested in a detailed discussion of the classification of the singularities of linear ODEs and of the related local analysis of solutions near the singular points is referred to [4.2]. As a simple example, let us consider the linear ODE

$$\begin{cases} \dfrac{dy}{dx} + \dfrac{y}{(1-x)^2} = 0, \\ y(0) = y_0, \end{cases} \tag{3.9a}$$

with solution

$$y(x) = y_0 \exp\left(\frac{x}{x-1}\right). \tag{3.9b}$$

The singularity occurs at $x = 1$, the pole of the coefficient p_0, and its location is *independent* of y_0, as expected. Let us now consider the non–linear ODE

$$\begin{cases} \dfrac{dy}{dx} - y^2 = 0, \\ y(0) = y_0, \end{cases} \tag{3.10a}$$

with solution

$$y(x) = \frac{y_0}{1 - x y_0}. \tag{3.10b}$$

The singularity occurs at $x = 1/y_0$, i.e., its location *depends* on y_0. These simple examples demonstrate that, in the N-body problem, it is natural to expect movable singularities in the particle orbits, which take place at times dependent on the assigned initial conditions.

A natural question arises. If we assume *good* initial conditions (i.e., absence of colliding particles and/or particles with infinite velocity at $t = 0$), can we say something on the time interval where we can exclude the occurrence of singularities in the N-body problem? An answer is provided by the following theorem (see, e.g., [2.1]):

Theorem 3.8 *For the ODE defined in eq. (2.1a), if $\exists\, l > 0$ so that for $|x_i - x_i^0| < l$ $(i = 1, ..., n)$*

I) $\mathbf{W} \in \mathcal{C}^\infty$

II) $\exists \, \mathcal{M} > 0$ so that $|W_i| < \mathcal{M}, \quad (i = 1, ..., n),$

then, for $|t| < l/(1+n)\mathcal{M}$, the solution $\boldsymbol{\Psi} = \boldsymbol{\Psi}(\mathbf{x}^0; t) \in \mathcal{C}^\infty$ with respect to the variable t.

In other words, the previous result identifies a *minimum* time interval where the solutions cannot present singularities. It is very interesting to evaluate the bounds thus posed for the generic N-body problem.

In order to apply the previous theorem we need to work with *dimensionless* equations, i.e., we need to eliminate the physical scales from the system given in eq. (1.3a). The initial conditions are the natural choice as *normalization constants*; accordingly, we start our exercise with the following:

Definition 3.9 *With the notation introduced in Definition 3.1, and for* $i, j = 1, ..., N$, *let*

$$\begin{cases} r^0 := \min_{i \neq j} r^0_{ij} > 0, \\ v^0 := \max_i ||\mathbf{v}^0_i||. \end{cases} \tag{3.11a}$$

We adopt as physical scales

$$\begin{cases} r_{\mathrm{n}} = r^0, \\ v_{\mathrm{n}} = \sqrt{GM/r_{\mathrm{n}}}, \\ t_{\mathrm{n}} = r_{\mathrm{n}}/v_{\mathrm{n}}, \end{cases} \tag{3.11b}$$

and we define the dimensionless *positions, velocities, and time as*

$$\begin{cases} \boldsymbol{\xi}_i := \mathbf{x}_i/r_{\mathrm{n}}, \\ \boldsymbol{\nu}_i := \mathbf{v}_i/v_{\mathrm{n}}, \\ \tau := t/t_{\mathrm{n}}, \end{cases} \tag{3.11c}$$

for $i = 1, ..., N$.

It is a trivial exercise to show that in the new variables eqs. (1.3ab), once written component by component, becomes

$$\begin{cases} \dot{\xi}_{i,k} = \nu_{i,k} \\ \dot{\nu}_{i,k} = -\dfrac{G t_{\mathrm{n}}}{v_{\mathrm{n}} r_{\mathrm{n}}^2} \displaystyle\sum_{j=1, j \neq i}^{N} \dfrac{m_j(\xi_{i,k} - \xi_{j,k})}{\xi_{ij}^3}, \end{cases} \tag{3.12a}$$

with $\xi_{ij} := ||\boldsymbol{\xi}_i - \boldsymbol{\xi}_j||$, $k = 1, 2, 3$ and $i = 1, ..., N$. Moreover, from eq. (3.11a) the (dimensionless) initial conditions satisfy the following inequalities:

$$\begin{cases} \xi_{ij}^0 = ||\boldsymbol{\xi}_i^0 - \boldsymbol{\xi}_j^0|| \geq 1, \quad \forall \, i, j = 1, ...N \,\& \, i \neq j; \\ ||\boldsymbol{\nu}_i^0|| \leq v^0/v_{\mathrm{n}}, \qquad \forall \, i = 1, ..., N. \end{cases} \tag{3.12b}$$

We estimate now the upper limit on τ as prescribed by Theorem 3.8. For $k = 1, 2, 3$ and $i = 1, ..., N$, let us consider the box defined by

$$\begin{cases} |\xi_{i,k} - \xi^0_{i,k}| < l & (\Rightarrow \|\boldsymbol{\xi}_i - \boldsymbol{\xi}^0_i\| < \sqrt{3}l); \\ |\nu_{i,k} - \nu^0_{i,k}| < l, \end{cases} \tag{3.13}$$

where for the moment l is unspecified. We now search for a majorant of the r.h.s. of eqs. (3.12a). The first step is given by the following

Lemma 3.10 *For* $l = \sqrt{3}/16$

$$\xi_{ij} > \frac{\xi^0_{ij}}{2}; \tag{3.14a}$$

$$|\xi_{i,k} - \xi_{j,k}| < \frac{3\xi^0_{ij}}{2}. \tag{3.14b}$$

PROOF   Inequality (3.14a) is proved as follows:

$$\begin{aligned} \xi^2_{ij} &= \|\boldsymbol{\xi}_i - \boldsymbol{\xi}^0_i + \boldsymbol{\xi}^0_i - \boldsymbol{\xi}_j + \boldsymbol{\xi}^0_j - \boldsymbol{\xi}^0_j\|^2 \\ &= (\xi^0_{ij})^2 + \|\boldsymbol{\xi}_i - \boldsymbol{\xi}^0_i + \boldsymbol{\xi}^0_j - \boldsymbol{\xi}_j\|^2 + 2 < \boldsymbol{\xi}_i - \boldsymbol{\xi}^0_i + \boldsymbol{\xi}^0_j - \boldsymbol{\xi}_j, \boldsymbol{\xi}^0_i - \boldsymbol{\xi}^0_j > \\ &\geq (\xi^0_{ij})^2 - 2| < \boldsymbol{\xi}_i - \boldsymbol{\xi}^0_i + \boldsymbol{\xi}^0_j - \boldsymbol{\xi}_j, \boldsymbol{\xi}^0_i - \boldsymbol{\xi}^0_j > | \\ &\geq (\xi^0_{ij})^2 - 2\xi^0_{ij}\|\boldsymbol{\xi}_i - \boldsymbol{\xi}^0_i + \boldsymbol{\xi}^0_j - \boldsymbol{\xi}_j\| \end{aligned}$$

where the last expression is obtained by means of the Cauchy–Schwarz inequality. From the triangular inequality

$$\xi^2_{ij} \geq (\xi^0_{ij})^2 - 2\xi^0_{ij}(\|\boldsymbol{\xi}_i - \boldsymbol{\xi}^0_i\| + \|\boldsymbol{\xi}^0_j - \boldsymbol{\xi}_j\|) \geq (\xi^0_{ij})^2 - 4\sqrt{3}l\xi^0_{ij},$$

where the last expression is obtained by using the first of eqs. (3.13). Moreover, from eq. (3.12b), $(\xi^0_{ij})^2 \geq \xi^0_{ij}$, and so $\xi^2_{ij} \geq (\xi^0_{ij})^2(1 - 4\sqrt{3}l)$. Assuming now $l = \sqrt{3}/16$, we obtain eq. (3.14a). Inequality (3.14b) is proved as follows:

$$\begin{aligned} |\xi_{i,k} - \xi_{j,k}| &= |\xi_{i,k} - \xi^0_{i,k} + \xi^0_{i,k} - \xi_{j,k} + \xi^0_{j,k} - \xi^0_{j,k}| \\ &\leq |\xi_{i,k} - \xi^0_{i,k}| + |\xi_{j,k} - \xi^0_{j,k}| + |\xi^0_{i,k} - \xi^0_{j,k}| < 2l + \xi^0_{ij}, \end{aligned}$$

where the last inequality derives from the first of eqs. (3.13) and the triangular inequality. Again, from eq. (3.12b) $|\xi_{i,k} - \xi_{j,k}| < (2l + 1)\xi^0_{ij} < (3/2)\xi^0_{ij}$ for $l = \sqrt{3}/16$. This completes the proof.   ◁

We can now derive the final

Theorem 3.11 *For $l = \sqrt{3}/16$, the vector field describing the N-body problem is C^∞, and the constant $\mathcal{M}$ is given by*

$$\mathcal{M} = \max(12, 1 + v^0/v_n).$$ (3.15)

For

$$|t| < \frac{t_n \sqrt{3}}{16(1 + 6N)\mathcal{M}}$$ (3.16)

the solution of the general N-body problem is analytic.

PROOF First, from eq. (3.14a) it follows that the denominators in eq. (3.12a) remain strictly positive inside the box defined by eq. (3.13), and so the vector field is guaranteed to be analytic, as required by Theorem 3.10. Second, for a generic acceleration component in eq. (3.12a), we have

$$\left| \frac{Gt_n}{v_n r_n^2} \sum_{j=1; j \neq i}^{N} \frac{m_j(\xi_{i,k} - \xi_{j,k})}{\xi_{ij}^3} \right| < 12 \frac{Gt_n}{v_n r_n^2} \sum_{j=1; j \neq i}^{N} \frac{m_j}{(\xi_{ij}^0)^2} \leq 12;$$

here we have used the triangular inequality, Lemma (3.12), eq. (3.12b), the obvious inequality $\sum_{j=1; j \neq i}^{N} m_j < M$, and finally eq. (3.11b). Third, a majorant for the velocity field is obtained as $|\nu_{i,k}| \leq |\nu_{i,k} - \nu_{i,k}^0| + |\nu_{i,k}^0| < l + ||\nu_i^0|| < 1 + v^0/v_n$. This completes the proof of eq. (3.15). Inequality (3.16) is finally proved by considering that the dimension of phase–space for an N-body system is $n = 6N$, and by using eq. (3.11c). ◁

As a first application of the estimate given in eqs. (3.15-16), let us consider a (standard) elliptical galaxy of total mass M, with N stars, each of mass M_* ($M = NM_*$), and effective radius R_e (see Chapter 11). For simplicity let us generously assume that half–mass is contained inside R_e, and that the *minimum* interparticle distance r^0 equals the *mean* interparticle distance, i.e.

$$\frac{N}{2} \frac{4\pi(r^0)^3}{3} = \frac{4\pi R_e^3}{3} \Rightarrow r^0 = r_n = R_e \left(\frac{2}{N} \right)^{1/3}.$$ (3.17a)

Moreover, let assume that v^0 is the *virial velocity* (eqs. [3.28ab]), which, for the present purposes, can be evaluated as

$$v^0 = \sqrt{\frac{GM}{R_e}}.$$ (3.17b)

Thus, from eq. (3.11b), the ratio

$$\frac{v^0}{v_n} = \sqrt{\frac{r^0}{R_e}} = \left(\frac{2}{N} \right)^{1/6}$$ (3.17c)

is a quantity smaller than unity for $N > 2$. Therefore, we have $\mathcal{M} = 12$. Finally, we find

$$t_n = \frac{r_n}{v_n} = \frac{(r^0)^{3/2}}{\sqrt{GM}}. \tag{3.17c}$$

Assuming now as characteristic physical values $M_* = M_\odot \simeq 1.989 \times 10^{33}$ g and $R_e = 5$ kpc, we find that $r_n \simeq 3 \times 10^{-4} R_e \simeq 1.5$ pc, and so $t_n \simeq 10^2$ yr. As a consequence, due to the presence in the denominator of eq. (3.16) of a term of the order of 10^{11}, $|t|$ turns out to be so small that it is totally useless in astrophysical applications!

With the following second very simple example, we show how the main limitation with the estimate given in eq. (3.16) is in fact due to the presence at denominator of the number N that characterizes the system. Let us consider a 2-body system where the two particles have the same mass m and are released at mutual distance r^0 with zero velocity. In this case the time of collision can be explicitly evaluated (do it as an exercise) to be

$$t_{\text{sing}} = \frac{(r^0)^{3/2}}{\sqrt{2Gm}} \frac{\pi}{2}. \tag{3.18a}$$

In this case $N = 2$ and $\mathcal{M} = 12$, and so the solution is certainly regular for

$$|t| < \frac{(r^0)^{3/2}}{\sqrt{2Gm}} \frac{\sqrt{3}}{2496}, \tag{3.18b}$$

a time shorter by of a factor $\simeq 10^4$ with respect to that on which the singularity is actually occurring! Note that the previous Theorem 3.11 gives a *finite* estimate even for problems where the singularities are *not* present, e.g., as in the case of a 2-body problem with non–zero angular momentum.

What can be said about the nature of the singularities? First, they can be classified as described by the following

Definition 3.12a *Three different singularities are exhibited by N-body systems:*

I) Collisions: $\lim_{t \to t_{\text{sing}}} r = 0$;

II) Pseudo-collisions: $\liminf_{t \to t_{\text{sing}}} r = 0$ & $\limsup_{t \to t_{\text{sing}}} r > 0$;

III) Blow–up: $\lim_{t \to t_{\text{sing}}} R = \infty$, *with* $t_{\text{sing}} < \infty$.

Second, the following results hold (see, e.g., [2.1], [2.5]):

Theorem 3.12b

I) For $N = 3$ all singularities are collisions.

II) If $I' = O(1)$ for $t \to t_{\text{sing}}$, then the singularity is a collision.

III) If the N particles are on a straight line, then all singularities are collisions.

IV) A singularity at t_{sing} is a collision if and only if $U = O[(t - t_{\text{sing}})^{-2/3}]$.

V) The set of initial conditions leading to a collision in a finite time has zero (Lebesgue) measure.

We conclude this short review on the singularities of the N-body problem by proving the beautiful

Theorem 3.13 [Weierstrass–Sundman] *The global collapse[3.2] of an N-body system may occur only on a finite amount of time (i.e., $t_{\text{sing}} < +\infty$) and only in the case $\mathbf{L}' = 0$.*

PROOF   The first step is the derivation of an important inequality involving the magnitude of the angular momentum. In a generic inertial reference system S_0,

$$||\mathbf{L}|| = ||\sum_{i=1}^{N} m_i \mathbf{x}_i \wedge \mathbf{v}_i|| \leq \sum_{i=1}^{N} m_i ||\mathbf{x}_i \wedge \mathbf{v}_i|| \leq \sum_{i=1}^{N} m_i ||\mathbf{x}_i|| \cdot ||\mathbf{v}_i||$$

$$= \sum_{i=1}^{N} \sqrt{m_i} ||\mathbf{x}_i|| \cdot \sqrt{m_i} ||\mathbf{v}_i|| \leq \sqrt{\sum_{i=1}^{N} m_i ||\mathbf{x}_i||^2} \sqrt{\sum_{i=1}^{N} m_i ||\mathbf{v}_i||^2},$$

where the last inequality derives from the Cauchy–Schwarz inequality. As a consequence, $||\mathbf{L}||^2 \leq 2IT$. Eliminating now the kinetic energy using the Lagrange–Jacobi identity given by eq. (3.5), we obtain the so–called *Sundman inequality*:

$$||\mathbf{L}||^2 \leq I(\ddot{I} - 2E).$$

We move now from the generic reference system S_0 to the barycentric inertial reference system S' of the N-body system. By definition of global collapse $\lim_{t \to t_{\text{sing}}} U = -\infty$, and so from the Lagrange–Jacobi identity $\lim_{t \to t_{\text{sing}}} \ddot{I}' = +\infty$. This proves that a global collapse must occur on a finite time. In fact, from $\lim_{t \to t_{\text{sing}}} \ddot{I}' = +\infty$ a time t_+ exists so that $t > t_+ \Rightarrow \ddot{I}' > 2$, i.e., $I'(t) > t^2 + \alpha t + \beta$: if $t_{\text{sing}} = +\infty$ then $\lim_{t \to t_{\text{sing}}} I' = +\infty$, against the hypothesis of global collapse.

[3.2] "Global collapse" means that all interparticle distances go simultaneously to zero, i.e., $\lim_{t \to t_{\text{sing}}} I'(t) = 0$ and $\lim_{t \to t_{\text{sing}}} U = -\infty$, where I' is given by eq. (3.6). See, e.g., [1.2].

It is straightforward to show that for $t \in [t_+, t_{\mathrm{sing}}]$ $\dot{I}' < 0$ and $\dot{I}'$ is a monotonically increasing function, i.e., $\dot{I}'(t) > \dot{I}'(t_+)$ and so $\dot{I}'(t_+)^2 - \dot{I}'(t)^2 \geq 0$. After multiplication of the Sundman inequality by the (positive) quantity $-\dot{I}'/I'$ and after integration over $[t_+, t]$ (with $t \leq t_{\mathrm{sing}}$), one obtains $||\mathbf{L}'||^2 \ln[I'(t_+)/I'(t)] \leq [\dot{I}'(t_+)^2 - \dot{I}'(t)^2]/2 + 2E'[I'(t) - I'(t_+)] \leq 2E'I'(t) + const.(t_+)$, i.e., $||\mathbf{L}'||^2 \leq [2E'I'(t) + const.(t_+)]/\ln[I'(t_+)/I'(t)]$, and for $t \to t_{\mathrm{sing}}$, $||\mathbf{L}'|| \to 0$. But the total angular momentum is conserved, and so $\mathbf{L}' = 0$.   ◁

3.3 Special solutions

Under particular circumstances (i.e., symmetries), the N-body problem can be solved explicitly. Euler and Lagrange explored the case of the 3-body problem, while the general problem is still actively investigated today. These solutions are of particular interest, because they are the only ones where the celebrated problem can be investigated in all details. In this subsection only a very short overview of known results will be given. The interested reader will find many more results (with proofs) in [1.2], [2.1], and [2.5]. All the following results are expressed in the barycentric (inertial) frame S'. Let us start with the basic

Definition 3.14 *In the barycentric inertial reference system S' a solution of the N-body problem is:*

I) planar, if there exists a plane Π with a time–independent orientation containing all the bodies at any time;

II) flat, if there exists a plane $\Pi(t)$ containing all the bodies at any time;

III) sygyzyal at time t_0, if at $t = t_0$ the N particles all lie on the same straight line;

IV) rectilinear, if there exists a straight line Λ with a time–independent orientation containing all the bodies at any time;

V) collinear, if there exists a straight line $\Lambda(t)$ containing all the bodies at any time;

VI) homographic, if there exists a scalar function $\lambda(t) \geq 0$ and an orthogonal matrix $\mathcal{R} \in SO(3)$ so that $\forall t$ the position of each particle is related to its initial position by

$$\mathbf{x}_i = \lambda(t)\mathcal{R}(t)\mathbf{x}_i^0; \tag{3.20}$$

VII) homotethic, if in eq. (3.20) $\mathcal{R}$ is the identity for all times;

VIII) of relative equilibrium, if in eq. (3.20) $\lambda(t) = 1$ for all times;

IX) a central configuration, *if there exists a scalar function $\sigma(t) \geq 0$ so that*

$$\frac{\partial U}{\partial \mathbf{x}_i} = \sigma(t) m_i \mathbf{x}_i; \tag{3.21}$$

$\forall t$ and $i = 1, ..., N$.

The following results are recorded without proof (see, e.g., [2.1] and [2.5]):

Theorem 3.15a
I) If the solution is planar and $\mathbf{L}' \neq 0$, the plane Π coincides with the invariable plane Π_{inv} defined in eq. (3.4).
II) A planar solution may exist even for $\mathbf{L}' = 0$.
III) If $\mathbf{L}' = 0$, then any flat solution is planar.
IV) Every planar solution is flat.
V) Not every flat solution is planar (as a trivial example, note that the solution of the general 3-body problem is always flat).
VI) From points III) and V), it follows that any solution of the 3-body problem with $\mathbf{L}' = 0$ is planar.
VII) If Π_{inv} exists, then any sygyzyal configuration lies on Π_{inv}.
VIII) Every rectilinear solution is collinear.
IX) Every collinear solution is flat.
X) Every collinear solution is planar [in fact, if $\mathbf{L}' \neq 0$, from point VII) the straight line $\Lambda(t)$ lies on Π_{inv} $\forall t$, and so the problem is planar: if $\mathbf{L}' = 0$, from points III) and IX), the solution must be planar].

It is important to note that the most interesting solutions from a physical point of view are certainly the homographic motion and the central configuration; for example, the celebrated collinear Euler–Lagrange and planar (equilateral) Lagrange solutions of the N-body problem are homographic. We record here, without proof, only the following

Theorem 3.15b [Laplace theorem] *Being a central configuration is the necessary and sufficient condition for homographic solutions of the N-body problem.*

3.4 Asymptotic behavior at large times

Many exact and asymptotic results are known for a general N-body system when $t \to \infty$. Such theorems are highly technical in nature, and will not be proved nor stated here. For a short summary of these results the interested reader is referred to [1.2], [2.1], [2.5]. Here only two theorems are presented, in order to illustrate the nature of such results.

Theorem 3.16 [Minimum and maximum interparticle distances]
In the general N-body problem,
I) For $E' < 0$ a positive constant A exists so that $r(t) \leq A \; \forall t$.
II) For $E' = 0$, $r(t) = O(t^{2/3})$ and a positive constant A exists so that $R(t) \geq At^{2/3}$ for $t \to \infty$.
III) For $E' > 0$, $r(t) = O(t)$ and a positive constant A exists so that $R(t) \geq At$ for $t \to \infty$.

PROOF   For simplicity we prove here only point I) and the second statement of point III). Point I) is trivial: if $E' < 0$, then $0 \leq T' = E' - U$, i.e., $V > |E'|$, and from the second of eq. (3.7a) $r(t)$ remains limited from above. The second statement of point III) is proved using the Lagrange–Jacobi identity. In fact, from $E' > 0$, $\ddot{I}' = 2(T' + E') \geq E'$ $\forall t$, and a simple integration gives $I' \geq at^2$ for $t \to \infty$ and some $a > 0$. Using eq. (3.7b) the result is proved.   ◁

Point I) means that when $E' < 0$ the minimum interparticle distance $r(t)$ remains bounded, while nothing can be said about the maximum interparticle distance $R(t)$ (i.e., at variance with the 2-body problem, particles may escape from an N-body system with total negative energy). Point II) means that $r(t)$ cannot grow faster than $t^{2/3}$ when $E' = 0$ (and may well remain bounded, of course); in contrast, $R(t)$ cannot grow slower than $t^{2/3}$, i.e., at least one particle must escape from an N-body system with vanishing total energy. From point III) it follows that, when $E' > 0$, $r(t)$ cannot grow faster than t (and may remain bounded), while $R(t)$ cannot grow slower than t: at least one particle must escape from an N-body system with total positive energy. Moreover, note that points II) and III) mean that $E' < 0$ is a *necessary* condition for all interparticle distances $r_{ij}(t)$ to be bounded at all times.

Before concluding the present Section, we introduce the basic concept of *virial*. This will be extensively investigated in the following Chapters.

Definition 3.17 [Virialized self–gravitating system] *An N-body system is called* self–gravitating *if the total potential under which the N particles move is determined by the masses themselves through eq. (1.3a). An N-body self–gravitating system is called* virialized *if in the barycentric inertial reference system S'*

$$2T' = -U > 0, \quad \forall t. \tag{3.22}$$

Theorem 3.18 *A self–gravitating N-body system is virialized if and only if*

$$\frac{dI'}{dt} = 0. \tag{3.23}$$

PROOF   The sufficient condition is a trivial application of the Lagrange–Jacobi identity: $\dot{I}' = 0 \Rightarrow \ddot{I}' = 0$, and so $2T' = -U$. The necessary condition is again proved using the Lagrange–Jacobi identity: in this case $\ddot{I}' = 0$, and so $I'(t) = \dot{I}'(0)t + I'(0)$. If $\dot{I}'(0) = 0$, eq. (3.23) is proved. If $\dot{I}'(0) \neq 0$, then for $t \to \infty$ or $t \to -\infty$ it follows that $I'(t) \to -\infty$, which is absurd.   ◁

Theorem 3.19 *If a self–gravitating N-body system is virialized, then*

$$E' < 0. \tag{3.24a}$$

Moreover, a self–gravitating N-body system is virialized if and only if

$$T' = T'(0); \qquad U = U(0); \tag{3.24b}$$

and two positive constants A and B exist such that

$$A \leq r_{ij}(t) \leq B \qquad (\forall i, j = 1, ..., N \,\&\, i \neq j). \tag{3.24c}$$

PROOF   From $2T' + U = 0$ and $T' + U = E'$ it follows that $0 \leq T' = -E'$ and $U = 2E'$, i.e., $E' \leq 0$; $E' = 0$ is discarded by definition, and so eq. (3.24a) is proved. In order to prove the necessary part of the Theorem, we assume that the system is virialized. From previous identities and total energy conservation, T' and U remains constant, as shown by eq. (3.24b). Now, $2|E'| = V \geq Gm_im_j/r_{ij}$, and $A = Gm_im_j/2|E'|$ in eq. (3.24c). From eq. (3.23), $I'(0) = I'(t) \geq m_im_jr_{ij}^2/M$ for $i,j = 1, ..., N$ and $i \neq j$, and so $B = \sqrt{I'(0)M/(m_im_j)}$. The sufficient part of the Theorem is proved as follows. If T' and U are constants, then integrating Lagrange–Jacobi identity we obtain $I'(t) = [2T'(0) + U(0)]t^2 + \dot{I}'(0)t + I'(0)$. Using eq. (3.24c) we immediately obtain that

I' is a bounded function, and this requires that the time coefficients in the expression for $I'(t)$ are zero, i.e., $I'(t) = I'(0)$. Theorem (3.18) proves that the system is virialized. ◁

A more general definition of virialized system is the following

Definition 3.20 *Let $A = A(t)$ a given function, so that $\forall t$*

$$< A >_t := \frac{1}{t} \int_0^t A(\tau) d\tau \tag{3.25}$$

exists. If $\lim_{t \to \infty} < A >_t$ exists, such limit will be indicated by $< A >_\infty$, and it is called time–average *of A.*

Theorem 3.21 [Time–averaged virial] *For a self–gravitating N-body system the following identities are equivalent*

$$2 < T' >_\infty = - < U >_\infty, \tag{3.26a}$$

$$\dot{I}'(t) = o(t), \qquad \text{for } t \to \infty. \tag{3.26b}$$

PROOF   The time–average operator of the Lagrange–Jacobi identity is

$$\frac{\dot{I}'(t) - \dot{I}'(0)}{t} = 2 < T' >_t + < U >_t,$$

and this proves the theorem.   ◁

Let us note that, if $\dot{I}'(t) = o(t)$ for $t \to \infty$, then $I'(t) = o(t^2)$ (asymptotic relations can always be integrated), and from eq. (3.7b) $R(t) = o(t)$ for $t \to \infty$. Also, if Theorem 3.21 holds, from the energy conservation and from the obvious fact that $< E >_t = E$, we have

$$< T' >_\infty = -E'; \qquad < U >_\infty = 2E'. \tag{3.27}$$

Definition 3.22 *For a virialized system, the* virial velocity dispersion V_V *and the* virial radius r_V *are defined as:*

$$\frac{MV_V^2}{2} := T', \qquad \frac{GM^2}{r_V} := |U|. \tag{3.28a}$$

In the new variables

$$V_V^2 = \frac{GM}{r_V}. \tag{3.28b}$$

4. THE LIOUVILLE APPROACH

This can be considered as a continuation of the previous Chapter, because it is still strictly based on the N-body problem. On the other hand, it is a natural transition to the following chapters where the methods and ideas at the basis of the continuum description of Stellar Dynamics will be developed. We first introduce three important concepts familiar in Statistical Mechanics, i.e., those of microstate, macrostate, and ensemble. Then we derive the so–called Liouville equation for the N-body problem. In addition, we provide a short description of quasi–linear partial differential equations and of the method of characteristics used for their solution. Finally, we show that the use of the method of characteristics to solve the Liouville equation is as difficult as solving the original set of ordinary differential equations associated with the N-body problem.

4.1 Basic concepts

From Chapter 3 it should be clear that the direct approach to the N-body problem (i.e., the search for a general solution of eq. [1.3a] followed by its investigation when specialized to stellar systems) is hopeless, and so new ways need to be explored. A possible idea is to restrict our interest to *some specific properties* of the N-body system, with the hope that the related evolutionary (differential) equations are easier to investigate. A particularly important family of functions, well suited for this approach, are the so–called *macroscopic functions* of an N-body system. A clear definition of macroscopic function is founded on two basic geometrical properties of phase–space and on the concepts of *microstate* and *macrostate*; the latter are borrowed directly from Statistical Mechanics (see, e.g., [1.4]).

In the following the phase–space of an N-body system $\Re^{6N}$ will be indicated with the standard name Γ. The extended phase–space is then $\Gamma \times \Re$, where the time coordinate appears. As a consequence, at any time the dynamical state of the N-body system is completely determined by a representative point in the extended phase–space. Second, the *evolution* of the system is a curve in the extended phase–space, originating from its *initial conditions*. The curve is the solution of eqs. (1.3ab). For the uniqueness of solutions, two different curves in the extended phase–space cannot intersect each other at any time.

Definition 4.1 [Microstate and macrostate] *Each point of the extended phase–space $\Gamma \times \Re$ is called a* microstate *for the N-body system. Any function Ξ defined on the extended phase–space is called* macrostate.

Note that in the previous definition the specific form (and nature as well, i.e., scalar, vectorial, etc.) of Ξ is left totally arbitrary, depending on the particular property of the system under investigation. Obviously, given an N-body system and its initial conditions, its trajectory in the extended phase–space can be interpreted as the set of all microstates where the system itself is found as a result of its specific initial conditions. For the uniqueness of solutions, the same system with different initial conditions can never be found with the same microstate. Moreover, as described in Chapter 2, the evolution of an assigned macrostate Ξ is given by its Lagrangian function $\Xi_{\mathcal{L}}$. The main question is: what is the mathematical property that any macrostate should have in order to "discard" some information from the full

N-body problem? Probably an intuitive answer to this (very complicated) problem is best approached by examples. Suppose we are interested in some *structural* property only of the system, i.e., we are not interested at all in the evolution of particle velocities. This means that in this case Ξ must be *invariant* with respect to all microstates that are identical in *configuration space*, but with generic conditions in *velocity space*. Another example can be obtained by considering a system for which all particles have the same mass, and for which we are not interested in the orbits of single particles: in this case all microstates obtained by permuting the phase–space are equivalent. Of course, in the "extreme case" where we ask for a complete description of the system (for example, using a macrostate equal to unity on the phase–space trajectory of the system, and zero otherwise), we are left with the initial N-body problem.

These simple considerations should convince us that, in general, a microstate defines unambiguously a macrostate, but the converse is not true. A given value for a prescribed macrostate determines in the extended phase–space many microstates.

From this point of view, the question we tried to formulate earlier now becomes: suppose that at $t = 0$ a system with specific initial conditions and a particular macrostate Ξ are assigned. Obviously, the value of Ξ is *completely determined* by the initial microstate. Can we predict the time evolution of Ξ with the assigned initial value? Is the related differential equation simpler than the original equations of the full N-body problem? After all, this was the strategy presented at the beginning of this Chapter. Note that, due to the invariance of Ξ with respect to some (or many) microstates, assigning an initial value for Ξ is equivalent to assign a set of microstates at $t = 0$. This can be better formulated by the following

Definition 4.2 [Ensemble] *The set of microstates that at $t = 0$ leave the macrostate Ξ invariant is called* ensemble (Ω_0) *for the macrostate Ξ.*

Obviously, each microstate $\in \Omega_0$ evolves following the differential equations of the N-body problem, and accordingly the ensemble Ω_0 becomes Ω_t. Note that Ξ was, by definition, invariant over Ω_0, but this in general cannot be expected to be true for Ω_t. This can be easily understood by considering that the evolution of coordinates in phase–space depends on *all* variables, and, even if the macrostate can be independent on some variable, also these "invisible" variables drive the evolution of the "visible" ones.

4.2 The Liouville equation

A possibility to proceed further with this approach is to define the *mean value* for the macrostate Ξ at $t \neq 0$ over Ω_t; this mean value is what was called *macroscopic function* in the introduction of the present Section. In order to define formally the concept of macroscopic function and to determine its evolution equation, we introduce the following

Theorem 4.3 [Liouville equation] *Let* $f^{(6N)}$ $: \Gamma \times \Re \mapsto \Re^+ \cup \{0\}$ *be the characteristic function for the ensemble* Ω_t, *i.e.,* $f^{(6N)}$ *equals unity if the microstate* $\in \Omega_t$, *zero otherwise. The time evolution of* $f^{(6N)}$ *is given by the solution of*

$$\frac{Df^{(6N)}}{Dt} = \frac{\partial f^{(6N)}}{\partial t} + \sum_{i=1}^{N} \left(< \mathbf{v}_i, \frac{\partial f^{(6N)}}{\partial \mathbf{x}_i} > - \frac{1}{m_i} < \frac{\partial U}{\partial \mathbf{x}_i}, \frac{\partial f^{(6N)}}{\partial \mathbf{v}_i} > \right) = 0,$$

(4.1a)

with the initial condition

$$f^{(6N)}(\Gamma, 0) = f_0^{(6N)},$$

(4.1b)

where $f_0^{(6N)}$ *is the characteristic function of* Ω_0, *and* $U = -(G/2) \sum_{i,j=1; i \neq j}^{N} m_i m_j / r_{ij}$.

PROOF Let $A(0) \in \Gamma$ be a generic volume of phase–space, and Ω_0 an ensemble. The number of microstates inside $A(t) \cap \Omega_t$ cannot change in the course of the evolution [for the uniqueness of solutions microstates cannot cross the boundary $\partial(A(t) \cap \Omega_t)$]. Moreover, the differential vector field associated with the N-body problem derived from U is solenoidal. As a consequence, the Transport theorem applied to the identity

$$\frac{d}{dt} \int_{\partial A(t) \cap \Omega_t} f^{(6N)} d\Gamma = 0$$

determines the differential eq. (4.1a). See, e.g., [5.3], [5.5], [5.7]. ◁

Note that from its definition $f^{(6N)}$ is a nowhere negative function over the extended phase space; moreover, $\forall A(t)$ in the extended phase space, $\int_{A(t)} f^{(6N)} d\Gamma$ can be interpreted as the "number" of microstates $\in A(t)$.

We are now in the position to formulate rigorously the concept of *macroscopic function*, introduced qualitatively in Section 4.1:

Definition 4.4 [Macroscopic function] *Let* Ξ *be a macrostate, and* Ω_0 *its ensemble at* $t = 0$. *At time* t *the* macroscopic function *associated to* Ξ *is* *given by*

$$< \Xi >_{\Omega(t)} := \frac{\int_{\Omega_t} f^{(6N)} \Xi d\Gamma}{\int_{\Omega_t} f^{(6N)} d\Gamma}. \tag{4.2a}$$

Equation above shows clearly that the time evolution of the macroscopic function $< \Xi >$ is known (at least in principle) provided that the time evolution of $f^{(6N)}$ is known. Note that, by definition, at $t = 0$

$$< \Xi >_{\Omega(0)} = \Xi, \tag{4.2b}$$

but this is not generally true for $t > 0$. In the next Section we discuss the possibility to solve the Liouville equation.

4.3 A solution for the Liouville equation?

A solution for the Liouville equation can formally be exhibited. In fact, the Liouville equation belongs to the following class of well–known Partial Differential Equations.

Definition 4.5 [Quasi–linear PDE] *Let* $\mathbf{x} \in \Re^n$. *A first–order partial differential equation (PDE) of the following form*

$$a_0(\mathbf{x}, y; t) \frac{\partial y}{\partial t} + \sum_{i=1}^{n} a_i(\mathbf{x}, y; t) \frac{\partial y}{\partial x_i} = 0 \tag{4.4a}$$

$$y(\mathbf{x}; 0) = h(\mathbf{x}), \quad \forall \mathbf{x} \in A \subseteq \Re^n, \tag{4.4b}$$

is called a quasi–linear *homogeneous PDE with boundary condition h. A solution* $y = y(\mathbf{x}; t)$ *is a function* $y : \Re^n \times \Re \mapsto \Re$, *so that eq. (4.4a) is identically verified, and* $y(\mathbf{x}; 0) = h(\mathbf{x})$. *If the coefficients* a_0, a_i $(i = 1, \ldots, n)$ *do not depend on* y, *the PDE is said to be* linear.

Assume that the solution y be known. The following is an important definition:

Definition 4.6 [Characteristics] *The* characteristic *of eq. (4.4a) associated with* $\mathbf{x}^0$ *is the solution (if it exists) of the following system of ODEs:*

$$\begin{cases} \dfrac{dx_i}{dt} = \dfrac{a_i[\mathbf{x}, y(\mathbf{x}^0; t); t]}{a_0[\mathbf{x}, y(\mathbf{x}^0; t); t]} \\ x_i(0) = x_i^0. \end{cases} \tag{4.5}$$

The characteristic will be indicated as $\mathbf{c} = \mathbf{c}(\mathbf{x}^0; t)$, *where* $\mathbf{c}(\mathbf{x}^0; 0) = \mathbf{x}^0$.

Then:

Theorem 4.7 *A function y is a solution of eqs. (4.4ab) if and only if y is constant on the associated characteristics, i.e.,*

$$y[\mathbf{c}(\mathbf{x}^0; t); t] = h(\mathbf{x}^0), \quad \forall \mathbf{x}^0. \tag{4.6}$$

PROOF   Let $\mathbf{x}^0$ be fixed, and g the solution of eqs. (4.4ab). Then from eq. (4.4a)

$$0 = a_0 \left(\frac{\partial y}{\partial t} + \sum_{i=1}^{n} \frac{a_i}{a_0} \frac{\partial y}{\partial x_i} \right).$$

In particular, along each characteristic $\mathbf{c}$,

$$0 = a_0 \left(\frac{\partial y}{\partial t} + \sum_{i=1}^{n} \frac{dx_i}{dt} \frac{\partial y}{\partial x_i} \right)_{\mathbf{x}=\mathbf{c}} = a_0 \frac{dy[\mathbf{c}(\mathbf{x}^0; t); t]}{dt},$$

and so y is constant along each characteristic. The proof of the sufficient condition is proved in an analogous way.   ◁

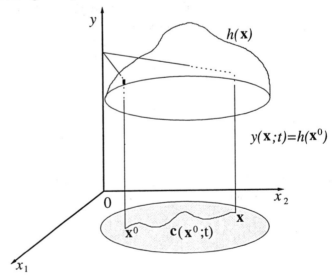

Figure 4.1

Note that, for a linear PDE, the characteristics are *independent* of the behaviour of the solution itself, and can thus be computed (at least in principle), without knowledge of the solution (see, e.g., [4.6]).

We are now in the position to discuss the characteristics of the Liouville equation. From eqs. (4.1ab) and (4.5) it appears that the characteristics of the Liouville equation are the solutions of the ODEs describing the motion of each particle in the N-body system: *despite its statistical nature, the Liouville approach is equivalent, in terms of complexity, to the direct solution of the original problem.* The reason why the Liouville approach fails to provide simpler equations describing the mean values of macrostates (i.e., the macroscopic functions) is due to the fact that these mean values are calculated over microstates (according to eq. [4.2a]), each of which evolves according to the complete set of ODEs of the N-body problem. The lesson is that – while maintaining a statistical approach – we should introduce additional assumptions on the behavior of N-body systems, i.e., we need to move from an exact *theory* to *modeling*.

5. THE 2-BODY RELAXATION TIME

How much of the star orbits in stellar systems depends on the forces acted by nearby stars? In this Chapter the problem of collisionality in stellar systems is presented in a simple approach. As an application of the framework developed here, the 2-body relaxation time is calculated in two idealized N-body systems. On the basis of the quantitative tools devised in this Chpater, we will show tht, broadly speaking, large stellar systems, such as elliptical galaxies, should be considered primarily as collisionless, while smaller systems, such as small globular clusters, may exhibit collisional behavior. Note that the behavior of N-body systems in collisionless and collisional regimes is very different, which is rich of astrophysical consequences, both from the observational and the theoretical point of view. In these Notes however we will mainly deal with collisionless stellar systems.

5.1 Basic concepts

As discussed in the previous Chapter, in order to extract information from the N-body problem, without actually solving it, we need to move to a different approach. In fact, the basic reason for the failure of the Liouville approach is that, despite its apparent statistical nature, the dimensionality of the phase–space Γ where the function $f^{(6N)}$ is defined, is the same as that of the original N-body problem. Suppose instead we find a way to replace $\Re^{6N}$ (the Γ phase–space) with $\Re^6$. Then we could reduce the N-body problem to the study of the motion of a *single* particle in the usual 3-dimensional space, i.e., in the *one–particle phase–space* γ. In this case the problem would be simplified enormously. Under suitable assumptions, such a reduction is indeed possible. The task of this Chapter is to present the physical basis for such a reduction, and then to derive the required mathematical formulation. With this reduction, we enter the field of the *model* approach to Stellar Dynamics.

In order to operate the dimensionality reduction we replace the *discrete* (discontinuous) distribution of the N mass particles in $\Re^3$ with a *smooth* (continuous) density distribution, $\rho(\mathbf{x};t)$. Nothing is specified about *how* such smoothing is performed. The density distribution ρ, taken to be a satisfactory description of the true "granular" N-body system, will serve as a *continuous model* for the original system. A smooth potential ϕ, naturally associated with ρ via the Poisson equation $\triangle\phi = 4\pi G\rho$, is given by

$$\phi(\mathbf{x};t) = -G \int \frac{\rho(\boldsymbol{\xi};t)d^3\boldsymbol{\xi}}{||\mathbf{x} - \boldsymbol{\xi}||}, \tag{5.1a}$$

(see, e.g., [4.4],[4.8],[4.9],[5.3]). As long as this continuous approximation holds, the motion of each point mass is determined only by the smooth (in general time–dependent) potential given by eq. (5.1a), and so the dimensionality of the relevant phase–space is reduced from $6N$ to 6. The main problem posed by the present approach is that in real systems each particle moves in the *true* potential ϕ_{true}, which for the individual particle i is given by

$$\phi_{\text{true}}(\mathbf{x}_i, t) = -G \sum_{j=1, j\neq i}^{N} \frac{m_j}{||\mathbf{x}_i(t) - \mathbf{x}_j(t)||}. \tag{5.1b}$$

The problem can thus be summarized in the following question: how large is the difference between the orbit described by a given particle under the influence of the force field derived from ϕ with respect to the true orbit described by ϕ_{true}? Or, stated differently, given to the fact that $\phi_{true} = \phi + (\phi_{true} - \phi)$, how long will it take to the granularity term $(\phi_{true} - \phi)$ to change significantly the orbit performed by each particle in the smooth potential ϕ? Of course, the "exact" answer presumes the knowledge of both orbits, and this is precisely the difficulty we would like to avoid.

In the next Section we will present the stellar dynamical approach to this problem, which will allow us to estimate the time scale over which the smooth description is physically reasonable, in a generic N-body system. Before presenting this discussion, we introduce a few basic definitions.

Definition 5.1 *For the N-body system the* nowhere negative *function* $n = n(\mathbf{x}, \mathbf{v}; t)$, $n : \gamma \times \Re \mapsto \Re^+ \cup 0$ *is defined by the property that*

$$\Delta N(\Delta \mathbf{x} \Delta \mathbf{v}; t) := \int_{\Delta \mathbf{x} \Delta \mathbf{v}} n(\mathbf{x}, \mathbf{v}; t) d^3 \mathbf{x} d^3 \mathbf{v} \qquad (5.2a)$$

is the number *of particles contained at time t in the 6-dimensional volume* $\Delta \mathbf{x} \Delta \mathbf{v} \subseteq \gamma$. *Obviously,*

$$\Delta N(\gamma; t) = \int_{\gamma} n(\mathbf{x}, \mathbf{v}; t) d^3 \mathbf{x} d^3 \mathbf{v} = N. \qquad (5.2b)$$

A function strictly related to n is the so–called

Definition 5.2 [Distribution function (DF)] *The* distribution function *(DF)* $f = f(\mathbf{x}, \mathbf{v}; t)$ *of an N-body system is a nowhere negative function* $f : \gamma \mapsto \Re^+ \cup 0$ *defined by the property that*

$$\Delta M(\Delta \mathbf{x} \Delta \mathbf{v}; t) := \int_{\Delta \mathbf{x} \Delta \mathbf{v}} f(\mathbf{x}, \mathbf{v}; t) d^3 \mathbf{x} d^3 \mathbf{v} \qquad (5.3a)$$

is the total mass *of the particles contained at time t in the 6-dimensional volume* $\Delta \mathbf{x} \Delta \mathbf{v} \subseteq \gamma$. *Obviously,*

$$\Delta M(\gamma; t) = \int_{\gamma} f(\mathbf{x}, \mathbf{v}; t) d^3 \mathbf{x} d^3 \mathbf{v} = M. \qquad (5.3b)$$

Note that in an N-body system with N_k particles of different masses m_k for $k = 1, ..., K$ (and so of total mass $M = \sum_{k=1}^{K} m_k N_k$), the functions n_k are naturally introduced following Definition 5.1; moreover,

$$n = \sum_{k=1}^{K} n_k, \qquad f = \sum_{k=1}^{K} m_k n_k. \tag{5.4}$$

A completely analogous set of definitions holds also for the smooth density distribution ρ. In this case the number density function n is not defined, but the distribution function is obtained from eq. (5.3a), now with the condition that

$$\rho(\mathbf{x}; t) = \int_{\Re^3} f(\mathbf{x}, \mathbf{v}; t) d^3\mathbf{v}; \tag{5.5a}$$

moreover, if the system consists of K different density distributions ρ_k,

$$f = \sum_{k=1}^{K} f_k, \qquad \rho_k(\mathbf{x}; t) = \int_{\Re^3} f_k(\mathbf{x}, \mathbf{v}; t) d^3\mathbf{v}, \qquad \rho = \sum_{k=1}^{K} \rho_k. \tag{5.5b}$$

5.2 The 2-body relaxation time

The standard approach to the calculation of the so called *relaxation time*, i.e., the characteristic time–scale beyond which the continuous approximation introduced in Section 5.1 is no longer valid, is based on two assumptions:

I) All encounters between particles are treated as *independent*, i.e., the effect of each encounter is simply *added* to that of the others,

II) All encounters between particles are treated as *hyperbolic 2-body encounters*.

For these reasons, the derived characteristic time is called *2-body relaxation time*, and indicated as t_{2b}. A gravitational system for which the cumulative effects of the encounters are negligible will be called *collisionless*, or, otherwise, *collisional*. Obviously, real N-body systems are *never*, strictly speaking, collisionless. Nevertheless, we will show that *exact models* based on the equations derived under the hypotheses of perfect non–collisionality, play an important role in Stellar Dynamics.

The general treatment of the evaluation of t_{2b} can be found in the monographies [5.5] and [5.8]. Here we follow a similar, but significantly simpler approach, rigorously justified by points I)-II) above.

Suppose that we describe, at time $t = 0$, an N-body system of total mass M in a barycentric inertial reference system S_0 by the function $n = n(\mathbf{x}^0, \mathbf{v}^0)$; the particles that make the system have masses m_f (where "f" stands for "field"). Moreover, suppose that a "test" mass m_t is placed at $t = 0$ on the z axis of S_0 at $z = z_t^0$, with initial velocity $\mathbf{v}_t^0 = (0, 0, v_t^0)$ directed along the z axis. In addition, let $\|\mathbf{v}_t^0\| = |v_t^0|$ be such that the total energy of each pair (m_f, m_t) is positive. As a result of the last assumption each encounter will be hyperbolic, and so the request of point (II) above is satisfied. Since the complete treatment of the 2-body problem can be found in many books (e.g., see [2.2], [2.4], [5.3], [5.5]), the required definitions and results are stated here without justification.

Definition 5.3 *For each pair (m_t, m_f), with velocities $(\mathbf{v}_t, \mathbf{v})$ and positions $(\mathbf{x}_t, \mathbf{x})$, let*

$$\mu := \frac{m_t m_f}{m_t + m_f}, \quad \mathbf{V} := \mathbf{v}_t - \mathbf{v}, \quad \mathbf{R} := \mathbf{x}_t - \mathbf{x}. \tag{5.6}$$

Here μ is the reduced mass, and $\mathbf{R}$ and $\mathbf{V}$ are the relative position and relative velocity, respectively.

Let $\mathbf{V}(\pm\infty)$ be the *asymptotic* relative velocity of the two particles in the 2-body hyperbolic motion (i.e., their relative velocity for $t \to \pm\infty$, respectively), $\mathbf{V}^0 = \mathbf{v}_t^0 - \mathbf{v}^0$ be their initial relative velocity ($t = 0$), and finally $\mathbf{L}' = \mu \mathbf{R}^0 \wedge \mathbf{V}^0$ be the total (constant) angular momentum in the inertial barycentric reference system of the pair, S'.

Definition 5.4 [Impact parameter] *If $\|\mathbf{V}(-\infty)\| > 0$, the* impact parameter *b is defined by*

$$b := \frac{\|\mathbf{L}'\|}{\mu \|\mathbf{V}(-\infty)\|}, \tag{5.7}$$

where $\mathbf{L}'$ is the angular momentum in S'.

The following results hold:

Lemma 5.5

I) In the 2-body hyperbolic motion,

$$\|\mathbf{V}(-\infty)\| = \|\mathbf{V}(\infty)\|, \tag{5.8a}$$

II) The angle 2ψ between $\mathbf{V}(-\infty)$ and $\mathbf{V}(\infty)$ is related to the other orbital parameters as

$$\cos\psi = \cfrac{1}{\sqrt{1 + \cfrac{b^2 \|\mathbf{V}(-\infty)\|^4}{G^2 (m_t + m_f)^2}}}, \qquad (5.8b)$$

III) In S_0 [where the initial conditions $(\mathbf{v}_t^0, \mathbf{v}^0)$ are given], the change of the kinetic (= total) energy of m_t over all its orbit is given by

$$\Delta E_t = \mu < \mathbf{V}_{\rm CM}, \mathbf{V}(\infty) - \mathbf{V}(-\infty) >, \qquad (5.8c)$$

where $\mathbf{V}_{\rm CM}$ is the barycentric velocity of the pair (m_t, m_f) [5.1].

PROOF   Points I)-II) are standard results for the 2-body hyperbolic motion. In order to prove point III), it is sufficient to recall that at any time $\mathbf{v}_t = \mathbf{V}_{\rm CM} + \mu \mathbf{V}/m_t$ and $\mathbf{v} = \mathbf{V}_{\rm CM} - \mu \mathbf{V}/m_f$, and that, at $t = \pm\infty$, the total energy reduces to the kinetic energy. Using the fact that $\mathbf{V}_{\rm CM}$ is constant and eq. (5.8a), the result is proved.   ◁

Now we are in the position to quantify the concept of the 2-body relaxation time $t_{\rm 2b}$. One possibility is to calculate separately, for each encounter, $|\Delta E_t|$ given by eq. (5.8c), and to compare the cumulative effect of the encounters to the initial kinetic energy of the test particle. After a number of encounters so that the cumulative energy exchange equals E_t^0 the system is taken to have spent one *relaxation time*. In other words, $t_{\rm 2b}$ is defined here as the time required for $\sum |\Delta E_t|$ to become equal to the initial (kinetic) energy of the star. We may argue that by then the test star, on the average, will have altered its original energy by about the same amount. Under the assumptions made earlier, an estimate of the total variation of E_t^0 in one crossing of the system is given by:

$$\sum |\Delta E_t| = \int_\gamma n(\mathbf{x}^0, \mathbf{v}^0) |\Delta E_t| d^3\mathbf{x}^0 d^3\mathbf{v}^0. \qquad (5.9a)$$

Now, we can define the 2-body relaxation time $t_{\rm 2b}$. If each crossing requires a *crossing time* $(t_{\rm cross})$, then $t_{\rm 2b}$ is given by

$$t_{\rm 2b} = n_{\rm 2b} \times t_{\rm cross}, \qquad (5.9b)$$

[5.1] The identity in eq. (5.8c) describes the so–called *slingshot effect*, widely used in interplanetary missions.

where, by definition, the number of crossings n_{2b} required for the "relaxation" of the test particle orbit is given by

$$\frac{n_{2b}\sum|\Delta E_t|}{E_t^0} := 1, \quad \text{i.e.} \quad n_{2b} = \frac{E_t^0}{\sum|\Delta E_t|}. \tag{5.9c}$$

In the following Section we will calculate such integral explicitly in two special cases, and we will show how the result does not depend significantly on the particular geometry of the system. The total number N of particles turns out to be the main parameter of the problem.

5.3 Two explicit cases

In this Section we specialize the previous general discussion to two idealized systems for which the calculations can be performed analytically. Despite their simplicity, the present cases are able to illustrate clearly the qualitative behavior of t_{2b} as a function of geometry, of the number density n, and of the total number of field particles.

We first consider a case where the initial velocities of the field particles, in the barycentric inertial reference system S_0, can be written as

$$\mathbf{v}^0 = (0, 0, v^0), \tag{5.10a}$$

i.e., n depends on the z component of $\mathbf{v}$ only. We then assume that the initial velocity of the test mass is the *asymptotic velocity* on the hyperbolic orbit ($z_t^0 = -\infty$). With this assumption we have

$$E_t^0 = \frac{m_t(v_t^0)^2}{2}, \quad \mathbf{V}_{\text{CM}} = \frac{(0, 0, m_t v_t^0 + m_f v^0)}{m_t + m_f}, \quad \mathbf{V}(-\infty) = (0, 0, \Delta v), \tag{5.10b}$$

where $\Delta v := v_t^0 - v^0$. As a result, for any pair (m_t, m_f) $\mathbf{V}_{\text{CM}}$ and $\mathbf{V}(-\infty)$ are parallel; moreover, from eq. (5.8a) we have

$$\|\mathbf{V}(\infty)\| = \|\mathbf{V}(-\infty)\| = |\Delta v|, \quad \mathbf{L}' = \mu \Delta v(-y^0, x^0, 0), \tag{5.10c}$$

and finally from eqs. (5.7), (5.10b), (5.10c)

$$b = \sqrt{(x^0)^2 + (y^0)^2}. \tag{5.10d}$$

From the definition of the orbital angle ψ (usually chosen to be zero at the minimum orbital separation), the *deflection angle*[5.2] between $\mathbf{V}(\infty)$ and

---

[5.2] Note that in the framework assumed in this Section, only field particles with $v^0 < v_t^0$ contribute to the deflection.

$\mathbf{V}(-\infty)$ (and $\mathbf{V}_{\mathrm{CM}}$ as well, because in the present case $\mathbf{V}(-\infty)$ and $\mathbf{V}_{\mathrm{CM}}$ are parallel) is $\pi - 2\psi$. Using eqs. (5.8ac), the first of eqs. (5.10c), and the trigonometric identity $\cos(2\psi) = 2\cos^2\psi - 1$, we have that

$$|\Delta E_t| = \frac{2\mu||\mathbf{V}_{\mathrm{CM}}||\,|\Delta v|}{1 + \dfrac{b^2 \Delta v^4}{G^2(m_t + m_f)^2}}. \tag{5.11}$$

The final assumption that characterizes the following two examples is that $n = n(\mathbf{x}^0)\delta(v^0)$, where δ is the Dirac delta function. In other words, all field particles are taken to be at rest at $t = 0$, and so $\Delta v = v_t^0$. Equation (5.2b) then becomes

$$\int_{\Re^3} n(\mathbf{x}^0) d^3\mathbf{x}^0 = N, \tag{5.12a}$$

and so eq. (5.9a) reduces to

$$\frac{\sum |\Delta E_t|}{E_t^0} = \frac{4\mu}{m_t + m_f} \int_{\Re^3} \frac{n(\mathbf{x}^0) d^3\mathbf{x}^0}{1 + \dfrac{b^2(v_t^0)^4}{G^2(m_t + m_f)^2}}. \tag{5.12b}$$

We are now ready to apply the formula derived above.

Example 5.6 [Axially symmetric n]
If n is axially symmetric around the z axis, so that $n = n(b, z^0)$, from eq. (5.12b) one obtains

$$\frac{\sum |\Delta E_t|}{E_t^0} = \frac{4\mu}{m_t + m_f} \int_{-\infty}^{\infty} dz^0 \int_0^{\infty} \frac{2\pi n(b, z^0) b\, db}{1 + \dfrac{b^2(v_t^0)^4}{G^2(m_t + m_f)^2}}, \tag{5.12c}$$

where the required integration limits are determined by the particular form of n. For a homogeneous cylinder particle distribution, of length L and radius R, we have $n = N/\pi R^2 L$. A straightforward integration gives

$$\frac{\sum |\Delta E_t|}{E_t^0} = \frac{4\mu}{m_t + m_f} \frac{N}{\Lambda^2} \ln(1 + \Lambda^2), \tag{5.13a}$$

where $\ln(1 + \Lambda^2)$ is the so-called *Coulomb logarithm*, and

$$\Lambda = \frac{R(v_t^0)^2}{G(m_t + m_f)}. \tag{5.13b}$$

In order to obtain a quantitative estimate from the obtained relation, we may refer to the case $m_t = m_f = m$, so that the first factor on the r.h.s. of eq. (5.13a) reduces to unity. From the Virial theorem in eqs. (3.28ab), if we use $v_t^0 = V_V$, $R = r_V$ and $M = Nm$, we obtain

$$\Lambda = \frac{N}{2}. \tag{5.13c}$$

As a result, from eq. (5.13a) the $N \to \infty$ asymptotic relation holds:

$$\frac{\sum |\Delta E_t|}{E_t^0} = \frac{4\ln(1 + N^2/4)}{N} \sim \frac{8\ln N}{N}. \tag{5.13d}$$

The formula above gives the asymptotic estimate of the change of energy of the test particle in one crossing of the cylindrical distribution. From the definition of t_{2b},

$$t_{2b} \sim \frac{N t_{\text{cross}}}{8\ln N}, \tag{5.13e}$$

where $t_{\text{cross}} \simeq L/v_t^0$ is the *crossing time*. Note that t_{2b} increases with *increasing N*.

A natural question is: how do these results depend on the particular geometry of the system and its internal distribution of particles n? In order to (partially) answer this important question, we now explore a different case.

Example 5.7 [Spherically symmetric n]
In this case, we assume a *spherically symmetric* density distribution for the field particles, and so $n = n(r)$. The natural change of variables in eq. (5.12b) is from Cartesian to spherical coordinates. Taking φ $(0 \le \varphi < 2\pi)$ as azimuthal angle around the z-axis, and ϑ $(0 \le \vartheta < \pi)$ the angle between r and the z-axis, one obtains $b = r\sin\vartheta$, and the formula analogous to eq. (5.12c) becomes:

$$\frac{\sum |\Delta E_t|}{E_t^0} = \frac{4\mu}{m_t + m_f} 2\pi \int_0^\pi \sin\vartheta d\vartheta \int_0^\infty \frac{n(r)r^2 dr}{1 + \frac{r^2\sin^2\vartheta(v_t^0)^4}{G^2(m_t + m_f)^2}} =$$

$$= \frac{4\mu}{m_t + m_f} 4\pi \int_0^1 d\tau \int_0^\infty \frac{n(r)r^2 dr}{1 + \frac{r^2(1 - \tau^2)(v_t^0)^4}{G^2(m_t + m_f)^2}}, \tag{5.14a}$$

where in the last expression we have introduced the variable $\tau = \cos\vartheta$. For a given form of $n(r)$ the previous integral is best evaluated by calculating first the integral over r, and then over τ. Formally the integral over τ can be evaluated independently of the particular form of $n(r)$, thus leading to a compact form for the energy variation:

$$\frac{\sum |\Delta E_t|}{E_t^0} = \frac{16\mu\pi}{m_t + m_f} \int_0^\infty \frac{\text{arcsinh}\lambda}{\lambda\sqrt{1+\lambda^2}} n(r)r^2 \, dr, \qquad (5.14b)$$

where

$$\lambda = \frac{r(v_t^0)^2}{G(m_t + m_f)}. \qquad (5.14c)$$

In particular, if we take $n = 3N/4\pi R^3$, i.e., consider that the field particles are distributed homogeneously inside a sphere of radius R, eq. (5.14b) becomes

$$\frac{\sum |\Delta E_t|}{E_t^0} = \frac{12\mu}{m_t + m_f} \frac{NS(\Lambda)}{\Lambda^3}, \qquad (5.15a)$$

where

$$S(\Lambda) = \int_0^\Lambda \frac{\lambda \, \text{arcsinh}\lambda}{\sqrt{1+\lambda^2}} d\lambda = \sqrt{1+\Lambda^2} \, \text{arcsinh}\Lambda - \Lambda \sim \Lambda \ln(2\Lambda). \qquad (5.15b)$$

In eq. (5.15b) Λ is the same as the quantity appearing in eq. (5.13b), and the asymptotic expansion holds for $\Lambda \to \infty$. Inserting the asymptotic relation given by eq. (5.15b) in eq. (5.15a), using $m_t = m_f = m$ and the virial relation in eq. (5.13c), we finally obtain

$$\frac{\sum |\Delta E_t|}{E_t^0} \sim \frac{12 \ln N}{N}, \qquad (5.15c)$$

i.e.,

$$t_{2b} \sim \frac{N t_{\text{cross}}}{12 \ln N}, \qquad (5.15d)$$

to be compared with eq. (5.13e). Therefore, the geometrical effect is not very important. As an exercise, the reader can try to use different functional forms for $n(r)$ and test how the results change.

As an astrophysical application, let us consider a typical elliptical galaxy, where $N \simeq 10^{11}$ and $t_{\text{cross}} \simeq 2 \times 10^8$ yrs: the characteristic 2-body relaxation time is then estimated to be of the order of $10^{6 \div 7}$ Gyrs. Therefore, elliptical galaxies are essentially *collisionless systems* over cosmological time scales. In contrast, for globular clusters, where $N \simeq 10^6$ and $t_{\text{cross}} \simeq 10^6$ yrs, $t_{2b} \simeq 5 \times 10^9$ yrs. For such systems the cumulative effects of 2-body encounters are then expected to be important on time scales shorter than their age. This conclusion is confirmed by observations and numerical simulations (see, e.g., [5.8]).

Three important conclusions can be derived from the previous results. The first is that, in the case of a *homogeneous infinite system* (i.e., $R \rightarrow \infty$ and n=constant), the integrals in eqs. (5.12c) and (5.14b) *diverge*, i.e., the cumulative effects of *distant encounters* are dominant: this is a direct manifestation of the *long–range* nature of the gravitational field. The second important consequence is that, by increasing the number of particles at *fixed* total mass, the effects of *granularity* of ϕ_{true} become important on longer time scales, i.e, the continuous approximation is better realized over *finite* times. The last comment concerns the observational fact that astrophysical systems such as galaxies and globular clusters are *very inhomogeneous*, i.e, the density of their central regions is many orders of magnitude larger than their *mean density* (see Chapter 9). Therefore, t_{2b} is expected to depend strongly on the position inside such systems.

6. THE COLLISIONLESS BOLTZMANN EQUATION (CBE)

In this Chapter the Collisionless Boltzmann equation (CBE) is derived from heuristic arguments. This equation, which applies in the limit of perfectly collisionless stellar systems, represents the starting point for many of the topics discussed in the following chapters. As anticipated in Chapter 5, the distinction between collisional and collisionless systems is not sharp, because the existence of significant collisionality effects depends on the time interval over which the system is studied. For this reason, one might wish to look for a simple modification of the CBE applicable to stellar systems in weakly collisional regimes. Such equation, here derived in a simple framework, is the so–called Fokker–Planck equation. It describes stellar systems where the cumulative effects of (weak) collisions are important.

6.1 The CBE

In the previous Chapter we have obtained indications that, by increasing the number of particles in a system dominated by gravitational forces, the collisionless approximation (i.e., the substitution of the true discrete system with the continuum approximation, and the consequent substantial reduction of the phase–space dimensionality) is better and better realized over longer and longer time–scales. In the ideal limit of $N \to \infty$ we expect the collisionless approximation to be valid for any time. For the sake of generality, let us assume that, in addition to the potential ϕ associated with the smooth density ρ (eq. [5.1a]), an *external potential* $\phi_{\mathrm{ext}} = \phi_{\mathrm{ext}}(\mathbf{x}; t)$ is considered, i.e., each element of γ moves under the action of the total potential $\phi_{\mathrm{T}} = \phi + \phi_{\mathrm{ext}}$ (when $\phi_{\mathrm{ext}} = 0$ the system is called *self–gravitating*).

The main goal is now to determine the differential equation in the collisionless regime for the evolution of the (smooth) distribution function f as given by eqs. (5.5ab). The answer is given by the following

Theorem 6.1 [Collisionless Boltzmann equation] *In the collisionless regime, the distribution function evolves according to:*

$$\frac{Df}{Dt} = \frac{\partial f}{\partial t} + < \mathbf{v}, \frac{\partial f}{\partial \mathbf{x}} > - < \frac{\partial \phi_{\mathrm{T}}}{\partial \mathbf{x}}, \frac{\partial f}{\partial \mathbf{v}} > = 0, \qquad (6.1a)$$

where

$$\phi_{\mathrm{ext}} = \phi_{\mathrm{ext}}(\mathbf{x}; t), \quad \triangle \phi(\mathbf{x}; t) = 4\pi G \int_{\Re^3} f d^3 \mathbf{v}, \qquad (6.1b)$$

with

$$f(\mathbf{x}, \mathbf{v}; 0) = f_0(\mathbf{x}, \mathbf{v}). \qquad (6.1c)$$

PROOF   In the collisionless approximation we can use the same mathematical arguments used in the derivation of the Liouville equation. In fact, for any arbitrary volume $A(t)$ in the one–particle phase–space γ, the mass contained in $A(t)$ at any time t is given by eq. (5.3a) (where f is the distribution function of the smooth density). Each point of this volume – and each point on its boundary $\partial A(t)$ – moves according to the vector field determined by the gradient of the potential ϕ_{T}: from the uniqueness of solutions of ODEs (Theorem 2.2), no orbits can leave this volume. In other words, the mass (of the smooth density) is conserved for any arbitrary volume and any arbitrary time. Finally, the vector field induced by ϕ_{T} is solenoidal, so that, using the Transport theorem in its formulation given by eq. (2.17b), we obtain the evolution equation (6.1a) for f. See, e.g., [5.3], [5.5], [5.6], [5.7], [5.8].　　　　　　　　　　　　　　　　◁

As in the case of the Liouville equation, for the CBE the solution can also be formally obtained in terms of its characteristics, given by the solutions of the system

$$\begin{cases} \dot{\mathbf{x}} = \mathbf{v} \\ \dot{\mathbf{v}} = -\dfrac{\partial \phi_{\mathrm{T}}}{\partial \mathbf{x}}, \end{cases} \tag{6.2}$$

with initial conditions $\mathbf{x}(0) = \mathbf{x}^0$ and $\mathbf{v}(0) = \mathbf{v}^0$. In other words, the characteristics associated with the CBE are curves in the $(6+1)$-dimensional one–particle phase–space $\gamma \times \Re$. As a consequence, the problem is now enormously simplified with respect to that posed by the solution of the Liouville equation, where the characteristics are $6N+1$ dimensional curves in the extended phase–space $\Gamma \times \Re$. Unfortunately, we do not yet have a complete under-standing of the properties of orbits in 3-dimensional time–dependent (or even time–independent!) potentials. Therefore, the problem posed by the general solution of eq. (6.1a) – despite its apparent simplicity – is still too difficult to be solved in general cases. In the following Chapters some of the techniques developed to extract information from the CBE will be described in detail.

Here it is important to point out the main differences between the CBE and the Liouville equation. Both equations are derived rigorously. Therefore, the following discussion does not deal with the *mathematical* formulation of the previous equations, but rather with their *ability to describe a given phys-ical system*. Form this point of view, the Liouville equation is the *exact* equation describing the motion of a general N-body system in phase–space Γ, and the CBE is the *exact* equation describing the motion of a perfectly collisionless system in phase–space γ. The key problem is the use of the CBE in the description of *real* N-body systems (i.e., systems with finite N). From the semi–quantitative arguments developed in Chapter 5, we can assume (at variance with the Liouville equation) that the CBE is a *model* equation that can be used to approximately describe gravitational systems with *finite N* on time–scales shorter than t_{2b}. An obvious question arises: is it possible to modify the CBE in a way that permits the description of systems with a finite number of particles over time–scales longer than t_{2b}, i.e., gravitational systems in the *collisional* regime? Before we embark in a qualitative descrip-tion of a possible generalization of the CBE in this direction, a discussion of what we should expect in the collisional regime is in order.

Suppose that we have as initial condition an N-body system, described in phase–space γ by the *discrete* DF given in eq. (5.5a) and the associated potential ϕ_{true}, and, in parallel, consider its smooth description as given by the *continuous* DF given in eq. (5.5b), with the associated pair (ρ, ϕ). As time increases, two different families of orbits originate from the same initial conditions in $\gamma \times \Re$: the orbits of the particles in the real system, described under the action of ϕ_{true}, and the orbits described by each mass element of the smooth density ρ under the action of ϕ. Note that the uniqueness theorem for solutions of ODEs applies to both families of orbits, but nothing prevents real particles (moving under the action of ϕ_{true}) from entering or leaving the arbitrary control volume $A(t)$ (which evolves under the action of ϕ). This means that, if after a sufficient amount of time we perform again the smoothing on the real system, the smoothed DF will be different from the smooth DF evolved according to the CBE. From a mathematical point of view, the fact that the number of particles of the true system can change inside the control volume $A(t)$ associated with the smooth potential ϕ, can be modeled by adding to the r.h.s. of the CBE a *source (or collision) term*. This collision operator is expected to depend on the smooth DF itself, i.e., we can write

$$\frac{Df}{Dt} = C[f], \qquad (6.3)$$

where $C[f]$ is, for the moment, an unspecified functional that should depend not only on the present value of f, but also on its previous history, i.e., on f over all the extend phase–space. The interesting question of how should one specify the form of $C[f]$ required to capture the long–term behavior of collisional gravitational systems goes beyond the goals of these Lectures. In the following Chapters we will focus on perfectly collisionless systems. However, because of its importance, in the next Section one possible approach to this particular problem is briefly described.

6.2 The Fokker–Planck equation

The so–called Fokker–Planck equation is derived under specific (and physically motivated) assumptions on the nature of the functional $C[f]$. Its derivation can be obtained rigorously (in the wide context of the so–called Stochastic Calculus, see, e.g., [3.4]), but because of its technical nature, it is not reported here. For simplicity, we follow here one of the considerably simpler approaches that can be found in the literature (see, e.g., [5.3], [5.7], [5.8]).

The first assumption is that the collision operator $C[f]$ at time t is completely determined by f *at the same time*, i.e., the previous evolution history of the DF is ignored[6.1]. Then, under this hypothesis we can write

Lemma 6.2 *At any time t,*

$$C[f] = \left(\frac{\partial f}{\partial t}\right)_{+} + \left(\frac{\partial f}{\partial t}\right)_{-}. \tag{6.4}$$

The sign $\pm$ in eq. (6.4) refers to particles that enter and leave the phase–space around $\mathbf{w}$ as a result of encounters.

A probability function $\Psi(\mathbf{w}, \Delta\mathbf{w}; t)$ is then introduced, [where $\mathbf{w} := (\mathbf{x}, \mathbf{v}) \in \gamma$], so that $\Psi(\mathbf{w}, \Delta\mathbf{w}; t)d^6\Delta\mathbf{w}$ is the *scattering rate* (at time t) of particles in $\mathbf{w}$ in the phase–space volume element $d^6\Delta\mathbf{w}$ around $\mathbf{w} + \Delta\mathbf{w}$. The two terms of the collision operator in eq. (6.4) are related to the scattering function by the following identities:

Lemma 6.3

$$\left(\frac{\partial f}{\partial t}\right)_{+} = \int_{\gamma} \Psi(\mathbf{w} - \Delta\mathbf{w}, \Delta\mathbf{w}; t)f(\mathbf{w} - \Delta\mathbf{w}; t)d^6\Delta\mathbf{w}, \tag{6.5a}$$

$$\left(\frac{\partial f}{\partial t}\right)_{-} = -\int_{\gamma} \Psi(\mathbf{w}, \Delta\mathbf{w}; t)f(\mathbf{w}; t)d^6\Delta\mathbf{w}. \tag{6.5b}$$

Equation (6.3), with the above collision terms, is called Master equation.

PROOF   Equations (6.5ab) are immediate consequence of the physical meaning of Ψ and f.   ◁

The second assumption (the Fokker–Planck approximation) is motivated by the fact that the main effects of the gravitational scattering are due to *weak* encounters (see Chapter 5), and so we can assume that the only significant contribution to the integrals (6.5ab) is obtained from a small volume around $\mathbf{w}$. In particular, by expanding up to the second order the terms containing $\mathbf{w} - \Delta\mathbf{w}$ in the integrand of eq. (6.5a), and by adding eq. (6.5b), we obtain

Definition 6.4 [Fokker–Planck equation]

$$C[f] = -\sum_{i=1}^{6} \frac{\partial f \, D(\Delta w_i)}{\partial w_i} + \frac{1}{2}\sum_{i,j=1}^{6} \frac{\partial^2 f \, D(\Delta w_i \Delta w_j)}{\partial w_i \partial w_j}, \tag{6.6a}$$

[6.1] In Statistics this is called a *Markov property*.

where the diffusion coefficients *are given by*

$$\begin{cases} D(\Delta w_i) = \int_\gamma \Delta w_i \Psi(\mathbf{w}, \Delta \mathbf{w}; t) d^6 \Delta \mathbf{w} \\ D(\Delta w_i \Delta w_j) = \int_\gamma \Delta w_i \Delta w_j \Psi(\mathbf{w}, \Delta \mathbf{w}; t) d^6 \Delta \mathbf{w}, \end{cases} \qquad (6.6b)$$

for $i, j = 1, ..., 6$.

A third assumption is usually made, the so–called *local approximation*,

$$\Psi(\mathbf{w}, \Delta \mathbf{w}; t) = \delta(\Delta \mathbf{x}) \, \Psi_\mathbf{v}(\mathbf{v}, \Delta \mathbf{v}; t), \qquad (6.7a)$$

in which one assumes that collisions change *only* the velocities of particles, while their positions remain unchanged. In this case the Fokker–Planck equation becomes

Definition 6.5 [Fokker–Planck equation in the local approximation]
In the local approximation,

$$C[f] = -\sum_{i=1}^{3} \frac{\partial f}{\partial v_i} \frac{D(\Delta v_i)}{} + \frac{1}{2} \sum_{i,j=1}^{3} \frac{\partial^2 f}{\partial v_i \partial v_j} \frac{D(\Delta v_i \Delta v_j)}{}. \qquad (6.7b)$$

The explicit evaluation of the diffusion coefficients in the local approximation for a system of N particles interacting with the gravitational force can be found in the literature (see, e.g., [5.3], [5.8]).

Before concluding this Chapter it is important to recall another viable approach to the derivation of the CBE, the so–called *BBGKY hierarchy*. With this method, the starting point is the Liouville equation, repeatedly integrated over the phase–space coordinates of all particles, except one. The equation obtained is similar to eq. (6.3), where now $C[f] = C[f^{6N}]$. Using the non–correlation hypothesis for $N \to \infty$, the collision term is shown to be zero, and so the CBE is derived (see, e.g., [5.3] for a short qualitative discussion).

7. THE JEANS EQUATIONS
AND
THE TENSOR VIRIAL THEOREM

In this Chapter we derive the (infinite) set of equations sometimes known as "Jeans equations", by considering velocity moments of the CBE. These equations are very important for a physically intuitive modeling of stellar systems. In fact, the natural domain of existence of the solution of the CBE (the distribution function) is the 7-dimensional extended one–particle phase space, while the (time–dependent) Jeans equations are defined over the usual (3+1)-dimensional space. Moreover, in most cases the observational data do not sample directly the velocity coordinates of phase–space. Thus, the importance of theoretical tools such as the Jeans equations, allowing us to interpret directly observable quantities, are particularly welcome. The physical meaning of the quantities entering the Jeans equations is then clarified by a detailed comparison with analogous equations used in Fluid Dynamics. Finally, by taking spatial moments of the Jeans equations over the configuration space, the Virial theorem in tensorial form (TVT) is derived.

7.1 The method of Moments

As discussed in Chapter 6, the general solution of the CBE (eqs. [6.1ac]) depends on the knowledge of the properties of orbits in general 3-dimensional (usually time dependent) potentials, a problem well beyond the actual possibilities of Mathematics. Therefore, various techniques have been developed in order to extract information from the CBE. Such methods can be broadly divided into the so–called *method of Moments* and the construction of *particular solutions for stationary systems*. Of course, in the case of stationary systems the two approaches can be used together.

In this Chapter we present the first method. The basic idea in the method of Moments is to look for differential equations simpler than the CBE, describing the relations between particular functions defined as *moments* of the DF over the velocity space (Jeans equations) and over the configuration space (Tensor and Scalar Virial theorem; TVT and SVT, respectively). All the results presented in this Chapter are derived in an inertial reference system S_0; moreover, the convention of sum over repeated indices is used. The starting point in deriving the Jeans equations is the following

Definition 7.1 *Let $F = F(\mathbf{x}, \mathbf{v}; t)$, $F : \gamma \times \Re \mapsto \Re$ be a* microscopic *function. The associated* macroscopic *function $\overline{F} = \overline{F}(\mathbf{x}; t)$, $\overline{F} : \Re^3 \times \Re \mapsto \Re$, is defined as*

$$\overline{F}(\mathbf{x}; t) := \frac{1}{\rho(\mathbf{x}; t)} \int_{\Re^3} F(\mathbf{x}, \mathbf{v}; t) f(\mathbf{x}, \mathbf{v}; t) d^3 \mathbf{v}, \qquad (7.1a)$$

where

$$\rho(\mathbf{x}; t) = \int_{\Re^3} f(\mathbf{x}, \mathbf{v}; t) d^3 \mathbf{v}. \qquad (7.1b)$$

From now on, a bar over a symbol will represent the operator given in eq. (7.1a). If F is independent of $\mathbf{v}$, then $F = \overline{F}$; obviously, we have $\overline{\overline{F}} = \overline{F}$. From a physical point of view $\overline{F}$ can be interpreted as the mean value of F over all the velocities of particles that at time t determine the density ρ at $\mathbf{x}$: in this context, f is the *probability density* used as weight function.

The starting point of the method of Moments is to derive the differential equation describing the evolution of $\overline{F}$. This equation is easily obtained from the identity

$$\int_{\Re^3} F \frac{Df}{Dt} d^3 \mathbf{v} = 0, \qquad (7.2)$$

a direct consequence of the CBE (eq. [6.1a]). We now prove the important

Theorem 7.2 *The following identity holds:*

$$\frac{\partial \rho \overline{F}}{\partial t} + \frac{\partial \rho \overline{F v_i}}{\partial x_i} = -\rho \frac{\partial \phi_{\rm T}}{\partial x_i} \overline{\frac{\partial F}{\partial v_i}} + \rho \overline{\frac{\partial F}{\partial t}} + \rho v_i \overline{\frac{\partial F}{\partial x_i}}, \qquad (7.3a)$$

where $\phi_{\rm T} = \phi + \phi_{\rm ext}$, *and*

$$\phi(\mathbf{x}; t) = -G \int_{\Re^3} \frac{\rho(\boldsymbol{\xi}; t)}{||\mathbf{x} - \boldsymbol{\xi}||} d^3\boldsymbol{\xi}, \quad \phi_{\rm ext}(\mathbf{x}; t) = -G \int_{\Re^3} \frac{\rho_{\rm ext}(\boldsymbol{\xi}; t)}{||\mathbf{x} - \boldsymbol{\xi}||} d^3\boldsymbol{\xi}. \quad (7.3b)$$

PROOF   Using the CBE, the first term in eq. (7.2) is

$$\int \frac{\partial f}{\partial t} F d^3\mathbf{v} = \int \left(\frac{\partial fF}{\partial t} - f\frac{\partial F}{\partial t} \right) d^3\mathbf{v} = \frac{\partial \rho \overline{F}}{\partial t} - \rho \overline{\frac{\partial F}{\partial t}}, \qquad [*]$$

because $\mathbf{v}$ is independent of t, and so the time derivative and integration over $\mathbf{v}$ can be interchanged. The second term is

$$\int v_i \frac{\partial f}{\partial x_i} F d^3\mathbf{v} = \int \left(\frac{\partial fF v_i}{\partial x_i} - f\frac{\partial F v_i}{\partial x_i} \right) d^3\mathbf{v} = \frac{\partial \rho \overline{F v_i}}{\partial x_i} - \rho v_i \overline{\frac{\partial F}{\partial x_i}}, \qquad [**]$$

because $\mathbf{v}$ is independent of $\mathbf{x}$, and so the spatial derivative and integration over $\mathbf{v}$ can be interchanged. Finally, the third and last term in eq. (7.2) is

$$\int \frac{\partial \phi_{\rm T}}{\partial x_i} \frac{\partial f}{\partial v_i} F d^3\mathbf{v} = \frac{\partial \phi_{\rm T}}{\partial x_i} \int \left(\frac{\partial fF}{\partial v_i} - f\frac{\partial F}{\partial v_i} \right) d^3\mathbf{v} = -\rho \frac{\partial \phi_{\rm T}}{\partial x_i} \overline{\frac{\partial F}{\partial v_i}}, \qquad [***]$$

because $\phi_{\rm T}$ is independent of $\mathbf{v}$ and $f \to 0$ for $||\mathbf{v}|| \to \infty$ (i.e., particles with infinite velocity are not allowed, and so the integration over all the velocity space of the exact differential with respect to v_i evaluates to 0). Equation (7.3a) is now proved by adding eqs. [*], [**], and [***], while eqs. (7.3b) are a direct consequence of potential theory (see, e.g., [4.8], [4.9]).   ◁

7.2 The Jeans equations

With the aid of the previous general result we are now able to derive the so–called Jeans equations. Before, it is necessary to introduce the following

Definition 7.3 [Velocity moments] *As particularly important macroscopic functions, we define for* $i, j = 1, 2, 3$

$$\overline{v_i}(\mathbf{x}; t) := \frac{1}{\rho(\mathbf{x}; t)} \int_{\Re^3} v_i f d^3\mathbf{v}, \qquad (7.4a)$$

$$\overline{v_i v_j}(\mathbf{x}; t) := \frac{1}{\rho(\mathbf{x}; t)} \int_{\Re^3} v_i v_j f d^3 \mathbf{v}, \qquad (7.4b)$$

$$\sigma_{ij}^2(\mathbf{x}; t) := \frac{1}{\rho(\mathbf{x}; t)} \int_{\Re^3} (v_i - \overline{v_i})(v_j - \overline{v_j}) f d^3 \mathbf{v}. \qquad (7.4c)$$

$\overline{v_i}$ are the components of the streaming velocity, and the (symmetric) tensor σ_{ij}^2 is the velocity dispersion tensor.

Obviously, higher order velocity moments can be defined by considering higher order products between the velocity components in eqs. (7.4bc): however, because of their importance in applications, we will focus here mainly on the second–order velocity moments introduced above. In particular, it is trivial to prove that

$$\sigma_{ij}^2 = \overline{v_i v_j} - \overline{v_i}\,\overline{v_j}, \quad (i, j = 1, 2, 3). \qquad (7.5a)$$

Note that σ_{ij}^2 can be negative for $i \neq j$. The reason of the square in the notation of the velocity dispersion tensor is the following: given the symmetry of σ_{ij}^2, an orthogonal (rotation) matrix (in general dependent on $\mathbf{x}$ and t) exists so that the velocity dispersion tensor in the new reference system is diagonal. Moreover, from the definition of σ_{ij}^2, it follows that the diagonal elements in this reference frame are positive, so that the velocity dispersion tensor is a *positive definite* symmetric tensor. This leads to a simple geometrical interpretation of σ_{ij}^2: at each point of a stellar system, it is possible to associate a *velocity dispersion ellipsoid*, defined by

$$\sigma^2 = \sigma_{ij}^2 n_i n_j, \quad (||\mathbf{n}|| = 1). \qquad (7.5b)$$

In general the orientation and the axial ratios of the velocity dispersion ellipsoid change from place to place and with time inside a system. A particular class of systems is described by the following

Definition 7.4 [Isotropic systems] *The velocity dispersion tensor is called isotropic if* $\forall \mathbf{x}$

$$\sigma_{ij}^2(\mathbf{x}; t) = \sigma^2(\mathbf{x}; t)\delta_{ij} \qquad (7.6)$$

i.e., the velocity dispersion ellipsoid is everywhere a sphere. If the velocity dispersion tensor is not isotropic, it is called anisotropic.

The Jeans equations are now derived (see, e.g., [5.3], [5.5], [5.6], [5.7]).

Theorem 7.5 [Jeans equations] *The following identities hold:*

$$\frac{\partial \rho}{\partial t} + \frac{\partial \rho \overline{v_i}}{\partial x_i} = \frac{D\rho}{Dt} + \rho \frac{\partial \overline{v_i}}{\partial x_i} = 0. \tag{7.7a}$$

$$\frac{\partial \rho \overline{v_i}}{\partial t} + \frac{\partial \rho \overline{v_i v_j}}{\partial x_j} = -\rho \frac{\partial \phi_{\rm T}}{\partial x_i}, \quad (i = 1, 2, 3). \tag{7.7b}$$

Equation (7.7b) can also be written as

$$\frac{\partial \overline{v_i}}{\partial t} + \overline{v_j} \frac{\partial \overline{v_i}}{\partial x_j} = \frac{D\overline{v_i}}{Dt} = -\frac{\partial \phi_{\rm T}}{\partial x_i} - \frac{1}{\rho} \frac{\partial \rho \sigma_{ij}^2}{\partial x_j}, \quad (i = 1, 2, 3), \tag{7.7c}$$

where[7.1] $D/Dt = \partial/\partial t + \overline{v_i} \partial/\partial x_i$.

PROOF   The proof is straightforward and is obtained by choosing $F = 1$ and $F = v_i$ in eq. (7.3a). Equations (7.7c) are obtained from eqs. (7.7b) with some manipulation by using identity (7.5a) and eq. (7.7a).   ◁

Remark 7.5a [Higher order Jeans equations]
The Jeans equations can be extended to high orders, by considering at each step $F = v_i v_j$, $F = v_i v_j v_k$, etc..

Remark 7.5b [Analogies and differences with Fluid Dynamics]
The reader with a basic knowledge of Fluid Dynamics will certainly recognize the striking similarities between the Jeans equations and the fluid dynamical equations of *continuity* and *momentum* in the presence of a gravitational field and viscosity, without source terms. The analogy is not due to chance. In fact, the equations of Fluid Dynamics can be derived by complete analogy using the Transport theorem with $\rho(\mathbf{x}; t)$ as basic function and $u_i(\mathbf{x}; t)$ ($i = 1, 2, 3$) as basic differential field (generally not solenoidal)[7.2]. In fact, for a fluid with the properties stated above, the continuity equation is derived by imposing the mass conservation in each volume $V(t) \subseteq \Re^3$

$$\frac{d}{dt} \int_{V(t)} \rho d^3 \mathbf{x} = 0, \tag{7.8a}$$

[7.1] Note that the material operator in the Jeans equations can be *rewritten* using the D/Dt operator, but the equations are not *derived* directly from the Transport theorem (see eq. [7.2]).

[7.2] A very clear and beautiful derivation and discussion of the basic equations of Fluid Dynamics can be found in, e.g., [3.1], [3.2], [3.3], [3.5].

from the Transport theorem

$$\frac{D\rho}{Dt} + \rho\frac{\partial u_i}{\partial x_i} = \frac{\partial \rho}{\partial t} + \frac{\partial \rho u_i}{\partial x_i} = 0. \tag{7.8b}$$

The i-th component of the momentum equation is obtained from the Second Law of Dynamics

$$\frac{d}{dt}\int_{V(t)} \rho u_i d^3\mathbf{x} = -\int_{V(t)} \rho\frac{\partial \phi_T}{\partial x_i} d^3\mathbf{x} + \int_{\partial V(t)} p_{ij}n_j d^2\mathbf{x}, \tag{7.9a}$$

where the two terms at the r.h.s of eq. (7.9a) describe the *volume* and *surface* forces, respectively. In particular, in the surface integral, $\mathbf{n}$ is the unit vector normal to the surface $\partial V(t)$, and

$$p_{ij}(\mathbf{x};t) = -p(\mathbf{x};t)\delta_{ij} + \tau_{ij} \tag{7.9b}$$

is the *stress tensor*. According to the *Cauchy second law* (see, e.g., [3.7]) p_{ij} is a symmetric tensor. In eq. (7.9b) $p(\mathbf{x};t)$ is the *thermodinamical pressure*, and τ_{ij} is the *shear stress* tensor[7.3]. Using the divergence theorem, the surface integral can be transformed into a volume integral,

$$\int_{\partial V(t)} p_{ij}n_j d^2\mathbf{x} = \int_{V(t)} \frac{\partial p_{ij}}{\partial x_j} d^3\mathbf{x} = -\int_{V(t)} \frac{\partial p}{\partial x_i} d^3\mathbf{x}, \tag{7.9c}$$

(where the last identity holds for inviscid fluids) and finally, using the Transport theorem and eq. (7.8b), one obtains

$$\frac{Du_i}{Dt} = \frac{\partial u_i}{\partial t} + u_j\frac{\partial u_i}{\partial x_j} = -\frac{\partial \phi_T}{\partial x_i} + \frac{1}{\rho}\frac{\partial p_{ij}}{\partial x_j}, \quad (i = 1, 2, 3). \tag{7.9d}$$

The similarity between eqs. (7.7a)–(7.8b) and eqs. (7.7c)–(7.9d) is complete, if we identify $\overline{v_i}$ with u_i ($i = 1, 2, 3$), and $-\rho\sigma_{ij}^2$ with p_{ij}. Note how an isotropic velocity dispersion tensor in a stellar system corresponds to an inviscid fluid. Can this similarity be extended to the energy equations? We

---

[7.3] For the so–called *Newtonian fluids*,

$$\tau_{ij} = \lambda\delta_{ij}\frac{\partial u_i}{\partial x_i} + \mu\left(\frac{\partial u_i}{\partial x_j} + \frac{\partial u_j}{\partial x_i}\right),$$

where μ and λ are the *first* and *second* viscosity coefficients, respectively (see, e.g., [3.6]).

answer this question by deriving first the energy equation for a fluid with
heat conduction. The continuum formulation of the First Law of Thermodynamics is

$$\frac{d}{dt}\int_{V(t)}\left(E+\frac{\rho||\mathbf{u}||^2}{2}\right)d^3\mathbf{x} = -\int_{V(t)}\rho u_i\frac{\partial\phi_T}{\partial x_i}d^3\mathbf{x}$$

$$+\int_{\partial V(t)}u_i p_{ij}n_j d^2\mathbf{x}$$

$$-\int_{\partial V(t)}h_i n_i d^2\mathbf{x}, \tag{7.9e}$$

where E is the *internal energy* per unit volume, and in the r.h.s. the work
per unit time associated with the volume and surface forces is considered.
The *heat conduction* is described by the *heat flux vector* $\mathbf{h}$, and the minus
sign in front of the surface integral is associated with the fact that $\mathbf{n}$ is
directed outward the control volume $V(t)$. From the Transport theorem,
with simple manipulations the material derivative of the kinetic energy term
can be considerably simplified using the continuity and momentum equations
(eqs. [7.8b] and [7.9d]):

$$\frac{DE}{Dt}+E\frac{\partial u_i}{\partial x_i}=\frac{\partial E}{\partial t}+\frac{\partial Eu_i}{\partial x_i}=p_{ij}\frac{\partial u_i}{\partial x_j}-\frac{\partial h_i}{\partial x_i}. \tag{7.9f}$$

As a consequence, the total energy equation can be reduced to an equation
involving the internal energy only.

    We now move to the corresponding equation for collisionless systems.
Note that, strictly speaking, no quantity is available similar to internal energy, as only kinetic energy is a meaningful concept. Thus, we focus on eq.
(7.3a) with $F=||\mathbf{v}||^2/2$. We start by proving the following

Lemma 7.5c *For* $F=||\mathbf{v}||^2/2$,

$$\frac{\partial F}{\partial v_i}=\overline{v_i}, \tag{7.9g}$$

$$\overline{F}=\frac{||\overline{\mathbf{v}}||^2}{2}+\frac{\text{Tr}(\sigma^2)}{2}, \tag{7.9h}$$

$$\overline{v_i F}=\chi_i+\sigma_{ij}^2\overline{v_j}+\overline{v_i}\left[\frac{\text{Tr}(\sigma^2)}{2}+\frac{||\overline{\mathbf{v}}||^2}{2}\right], \tag{7.9i}$$

where

$$\chi_i := \frac{\overline{(v_i - \overline{v_i})||\mathbf{v} - \overline{\mathbf{v}}||^2}}{2}. \qquad (7.9j)$$

PROOF   The proof of eqs. (7.9gh) is trivial. The proof of eq. (7.9i) is obtained by expanding the r.h.s. of eq. (7.9j), and by using eq. (7.5a).   ◁

Equation (7.9h) suggests that, for a stellar system, the quantity $\rho \mathrm{Tr}(\sigma^2)/2$ be identified with the internal energy of Fluid Dynamics. After substitution of eqs. (7.9ghi) in eq. (7.3a), by working on the terms containing the kinetic energy $\rho||\overline{\mathbf{v}}||^2/2$, and using eqs. (7.7a) and (7.7c), one finally obtains

$$\frac{\partial E}{\partial t} + \frac{\partial E \overline{v_i}}{\partial x_i} = \frac{DE}{Dt} + E \frac{\partial \overline{v_i}}{\partial x_i} = -\rho \sigma_{ij}^2 \frac{\partial \overline{v_j}}{\partial x_i} - \frac{\partial \rho \chi_i}{\partial x_i}, \qquad (7.9k)$$

where the analogy with eq. (7.9f) is complete if we identify $\rho \chi_i$ with the heat conduction vector h_i.

Despite this striking formal analogy, a fundamental difference between fluids and stellar systems is apparent. In fact, while the equations of Fluid Dynamics are a *closed* set of equations because the pressure is related to ρ and E trough the *equation of state*, the concept of equation of state cannot be introduced from basic physical arguments in the Jeans equations. This means that the latter are an *infinite set* of equations. Thus, in order to solve the Jeans equations, *we are forced to assume a more or less arbitrary closure relation.* This is often done by specifying the properties of the velocity dispersion tensor. This aspect of the problem will be extensively discussed in the following Chapters.

7.3 The Tensor Virial theorem

It is possible to obtain additional relations between global functions associated with a stellar system, starting from the Jeans equations. The final result is a generalization of the Scalar Virial theorem (SVT), derived in Chapter 3. The Tensor Virial theorem (TVT) is now obtained by integrating the Jeans equations on the configuration space (see, e.g., [4.4] and [5.3]). The main quantities entering the TVT are now introduced.

Definition 7.6 [Inertia tensor] *For* $i, j = 1, 2, 3$,

$$I_{ij}(t) := \int_{\Re^3} \rho x_i x_j d^3\mathbf{x} \qquad (7.10a)$$

is the inertia tensor[7.4]. *Its trace is indicated by*

$$I(t) = \mathrm{Tr}(I_{ij}).$$
(7.10b)

The following tensors can be associated with the kinetic energy of the component described by the density ρ

Definition 7.7 [Kinetic energy tensors] *For* $i, j = 1, 2, 3$

$$K_{ij}(t) := \frac{1}{2} \int_{\Re^3} \rho \overline{v_i v_j} d^3 \mathbf{x}$$
(7.11a)

$$T_{ij}(t) := \frac{1}{2} \int_{\Re^3} \rho \overline{v_i}\, \overline{v_j} d^3 \mathbf{x}$$
(7.11b)

$$\Pi_{ij}(t) := \int_{\Re^3} \rho \sigma_{ij}^2 d^3 \mathbf{x}$$
(7.11c)

are the total, streaming, *and* dispersion kinetic energy tensors, *respectively. The trace of the above tensors is indicated respectively by*

$$K(t) = \mathrm{Tr}(K_{ij}), \quad T(t) = \mathrm{Tr}(T_{ij}), \quad \Pi(t) = \mathrm{Tr}(\Pi_{ij}).$$
(7.11d)

Two important consequences of the previous definitions, to be used in the following discussions, are straightforward:

Lemma 7.8 *For* $i, j = 1, 2, 3$,

$$I_{ij} = I_{ji}, \quad K_{ij} = K_{ji}, \quad T_{ij} = T_{ji}, \quad \Pi_{ij} = \Pi_{ji},$$
(7.12a)

$$K_{ij} = T_{ij} + \frac{\Pi_{ij}}{2}, \quad K = T + \frac{\Pi}{2}.$$
(7.12b)

[7.4] Note that the I_{ij} is related to the inertia tensor $\mathcal{I}_{ij}$ appearing in the expression of the kinetic energy of systems of points under a reference frame change (see, e.g., [1.1], [1.3], [1.5], [1.6], [1.7]), and to the quadrupole moment tensor Q_{ij} (see, e.g., [4.8]), by the relations

$$\mathcal{I}_{ij} = \mathrm{Tr}(I_{ij})\delta_{ij} - I_{ij}, \quad Q_{ij} = 3I_{ij} - \mathrm{Tr}(I_{ij})\delta_{ij}.$$

Also, note that I_{ij}, $\mathcal{I}_{ij}$, and Q_{ij} have the same eigenvectors and so are diagonal in the same reference frame.

Two scalar functions describe the gravitational energy of the density distribution ρ in the total potential ϕ_T:

Definition 7.9 [Self–energy and external energy] *For the density distribution ρ the two quantities*

$$U(t) = \frac{1}{2} \int_{\Re^3} \rho\phi d^3\mathbf{x} = -\frac{G}{2} \int_{\Re^6} \frac{\rho(\mathbf{x};t)\rho(\boldsymbol{\xi};t)}{||\mathbf{x} - \boldsymbol{\xi}||} d^3\mathbf{x}d^3\boldsymbol{\xi} \qquad (7.13a)$$

$$U_{\mathrm{ext}}(t) = \int_{\Re^3} \rho\phi_{\mathrm{ext}} d^3\mathbf{x} = -G \int_{\Re^6} \frac{\rho(\mathbf{x};t)\rho_{\mathrm{ext}}(\boldsymbol{\xi};t)}{||\mathbf{x} - \boldsymbol{\xi}||} d^3\mathbf{x}d^3\boldsymbol{\xi}, \qquad (7.13b)$$

are the gravitational self–energy[7.5] *and the* external energy, *respectively.*

Two tensors can be associated with the interaction between ρ and $\phi_T = \phi + \phi_{\mathrm{ext}}$:

Definition 7.10 [Self–energy and interaction energy tensors] *For $i, j = 1, 2, 3$*

$$U_{ij}(t) := -\int_{\Re^3} \rho x_i \frac{\partial\phi}{\partial x_j} d^3\mathbf{x}, \qquad (7.14a)$$

$$W_{ij}(t) := -\int_{\Re^3} \rho x_i \frac{\partial\phi_{\mathrm{ext}}}{\partial x_j} d^3\mathbf{x}. \qquad (7.14b)$$

are the gravitational self–energy *and* interaction energy *tensors, respectively.*

The following result holds:

[7.5] Note that the following identity holds:

$$\phi\triangle\phi = \frac{\partial}{\partial x_i}\left(\phi\frac{\partial\phi}{\partial x_i}\right) - \frac{\partial\phi}{\partial x_i}\frac{\partial\phi}{\partial x_i}.$$

If $\rho > 0$ over a volume $V \subseteq \Re^3$ bounded by ∂V with normal $\mathbf{n}$, then using the divergence theorem eq. (7.13a) can be rewritten as

$$U(t) = -\frac{1}{8\pi G} \int_V ||\nabla\phi||^2 d^3\mathbf{x} + \frac{1}{8\pi G} \int_{\partial V} \phi < \nabla\phi, \mathbf{n} > d^2\mathbf{x}.$$

If ∂V is an equipotential surface, then from the divergence theorem and the Poisson equation the surface integral in the identity above equals $\phi(\partial V)M/2$, where M is the system total mass. Finally, if $V = \Re^3$ and $\phi \to 0$ for $||\mathbf{x}|| \to \infty$, then the surface integral vanishes.

Theorem 7.11 *For $i, j = 1, 2, 3$*

$$U_{ij} = -\frac{G}{2} \int_{\Re^6} \frac{(x_i - \xi_i)(x_j - \xi_j)\rho(\mathbf{x};t)\rho(\boldsymbol{\xi};t)}{||\mathbf{x} - \boldsymbol{\xi}||^3} d^3\mathbf{x} d^3\boldsymbol{\xi}, \tag{7.15a}$$

$$U_{ij} = U_{ji}, \tag{7.15b}$$

$$U(t) = \mathrm{Tr}(U_{ij}), \tag{7.15c}$$

$$W(t) = \mathrm{Tr}(W_{ij}) = -\int_{\Re^3} \rho < \mathbf{x}, \frac{\partial \phi_{\mathrm{ext}}}{\partial \mathbf{x}} > d^3\mathbf{x}. \tag{7.15d}$$

PROOF   Equation (7.15b) is a direct consequence of eq. (7.15a); eq. (7.15c) is derived from eqs. (7.15a) and (7.13a); and finally eq. (7.15d) is a direct consequence of eq. (7.14b). In order to prove eq. (7.15a), from eqs. (7.14a) and the first of eqs. (7.3b), we note that

$$U_{ij} = -G \int \frac{x_i(x_j - \xi_j)\rho(\mathbf{x};t)\rho(\boldsymbol{\xi};t)}{||\mathbf{x} - \boldsymbol{\xi}||^3} d^3\mathbf{x} d^3\boldsymbol{\xi},$$

for $i, j = 1, 2, 3$. In the above equation $\mathbf{x}$ and $\boldsymbol{\xi}$ can be exchanged without affecting the value of U_{ij}, so that

$$U_{ij} = G \int \frac{\xi_i(x_j - \xi_j)\rho(\mathbf{x};t)\rho(\boldsymbol{\xi};t)}{||\mathbf{x} - \boldsymbol{\xi}||^3} d^3\mathbf{x} d^3\boldsymbol{\xi} :$$

eq. (7.15a) is then proved by adding the two above identities.   ◁

Note that the symmetry of U_{ij} is not evident from its definition given by eq. (7.14a). In contrast, in general W_{ij} is *not* a symmetric tensor. In order to derive the TVT we need to first prove the following:

Lemma 7.12 *For $i, j = 1, 2, 3$,*

$$\frac{dI_{ij}}{dt} = \int_{\Re^3} \rho \left(x_i\overline{v_j} + x_j\overline{v_i} \right) d^3\mathbf{x}. \tag{7.16}$$

PROOF   From eq. (7.10a)

$$\frac{dI_{ij}}{dt} = \int \frac{\partial\rho}{\partial t} x_i x_j d^3\mathbf{x} = -\int \frac{\partial\rho\overline{v_k}}{\partial x_k} x_i x_j d^3\mathbf{x} = \int \rho \left(x_i\overline{v_j} + x_j\overline{v_i} \right) d^3\mathbf{x},$$

for $i, j = 1, 2, 3$. The second identity in the above equation is obtained by using eq. (7.7a), and the last identity by integrating by parts over x_k.   ◁

We are now ready to prove the following

Theorem 7.13 [TVT and SVT] *For $i, j = 1, 2, 3$,*

$$\frac{1}{2}\frac{d^2 I_{ij}}{dt^2} = 2K_{ij} + U_{ij} + \frac{W_{ij} + W_{ji}}{2}, \qquad (7.17a)$$

$$\frac{1}{2}\frac{d^2 I}{dt^2} = 2K + U + W. \qquad (7.17b)$$

The first identity is the Tensor Virial theorem, *and the second identity is the* Scalar Virial theorem.

PROOF   Equation (7.17b) is a direct consequence of eqs. (7.17a) and (7.15d). The proof of identity (7.17a) is obtained by integrating over the configuration space the i-th component of eq. (7.7b) after multiplication by x_k, as follows. Let us define the resulting identity with the expression $A_{ki} = A_{ki}^{(1)} + A_{ki}^{(2)} - A_{ki}^{(3)} = 0$. The first term is given by

$$A_{ki}^{(1)} = \int x_k \frac{\partial \rho \overline{v_i}}{\partial t} d^3\mathbf{x} = \frac{d}{dt}\int \rho \overline{v_i} x_k d^3\mathbf{x}.$$

The second term is

$$A_{ki}^{(2)} = \int x_k \frac{\partial \rho \overline{v_i v_j}}{\partial x_j} d^3\mathbf{x} = -\int \rho \overline{v_i v_k} d^3\mathbf{x} = -2K_{ki},$$

where the first identity is obtained by integrating by parts with respect to x_j. The r.h.s. of eq. (7.7b) can be split into two contributions, corresponding to the fact that $\phi_T = \phi + \phi_{\text{ext}}$,

$$A_{ki}^{(3)} = -\int x_k \rho \frac{\partial \phi_T}{\partial x_i} d^3\mathbf{x} = U_{ki} + W_{ki}.$$

We now symmetrize the tensorial identity $A_{ki} = 0$ as $(A_{ki} + A_{ik})/2 = 0$: from the symmetry of U_{ij} and K_{ij}, and from eq. (7.16), the theorem is proved.   ◁

Note that a similar tensorial identity can be obtained when the system is made of N point masses, following the same approach used in the derivation of the Lagrange–Jacobi identity (eq. [3.5]). The only difference in the discrete case, with respect to the result obtained by using the Jeans equations, is that $K_{ij} = T_{ij}$, i.e., the dispersion kinetic energy tensor is obviously not defined. This means that the TVT and the SVT for multi–component systems are valid for a general N-body system interacting with an external potential, *independently* of any assumption about collisionality. In other words, the assumption of non–collisionality used in this Chapter *preserves* the correct TVT and SVT[7.6].

Particular attention should be payed to avoid an erroneous (and unfortunately frequent) application of the SVT to multi–component systems. In fact, let us assume that the system is made of the superposition of two density–potential pairs, namely (ρ_1, ϕ_1) and (ρ_2, ϕ_2). As a consequence, the total system is characterized by $\rho_T = \rho_1 + \rho_2$ and $\phi_T = \phi_1 + \phi_2$, and it is self–gravitating, in the sense that $\phi_{\rm ext} = 0$. Due to the additive nature of kinetic energy and inertia tensors, the total kinetic energy and the total moment of inertia are simply given by $K = K_1 + K_2$ and $I = I_1 + I_2$. From eq. (7.17b), the SVT for the total system is then

$$\ddot{I} = 2K + U, \tag{7.18a}$$

where

$$U = \frac{1}{2}\int_{\Re^3}\rho_T\phi_T d^3\mathbf{x} = \frac{1}{2}\int_{\Re^3}(\rho_1\phi_1 + \rho_2\phi_2 + \rho_1\phi_2 + \rho_2\phi_1)d^3\mathbf{x} =$$
$$= U_1 + U_2 + U_{1,2} = U_1 + U_2 + U_{2,1}. \tag{7.18b}$$

In the last two terms, the quantities

$$U_{1,2} := \int_{\Re^3}\rho_1\phi_2 d^3\mathbf{x}; \quad U_{2,1} := \int_{\Re^3}\rho_2\phi_1 d^3\mathbf{x}, \tag{7.18c}$$

[7.6] The interested reader can show, as a useful exercise, the following result. Under the same assumptions as in Theorem 7.13, if we further require that ρ vanishes outside the (time independent) regular volume V, with boundary ∂V and unit normal $\mathbf{n}$, then, if $n_i\overline{v_i} = 0$ on ∂V,

$$\ddot{I} = 2K + U + W - \mathcal{P},$$

where $\mathcal{P} := \mathrm{Tr}(\mathcal{P}_{ij})$ and

$$\mathcal{P}_{ij} = \int_{\partial V}\rho\sigma_{ij}^2 x_j n_k d^2\mathbf{x}.$$

If, in addition, $\rho\sigma_{ij}^2$ is constant on ∂V for $i, j = 1, 2, 3$, then from the divergence theorem

$$\mathcal{P} = V[\rho\mathrm{Tr}(\sigma_{ij}^2)]_{\partial V}.$$

A simple description of an important application of this result can be found in Section 8.2 of [5.3].

are the interaction energies of the two density distributions. We also have $U_{1,2} = U_{2,1}$ for the well–known *reciprocity theorem*, which follows immediately from eqs. (7.3b). The *correct* application of the SVT to each component is $\ddot{I}_1/2 = 2K_1 + U_1 + W_{1,2}$ and $\ddot{I}_2/2 = 2K_2 + U_2 + W_{2,1}$, where,

$$W_{1,2} = -\int_{\Re^3} \rho_1 < \mathbf{x}, \frac{\partial \phi_2}{\partial \mathbf{x}} > d^3\mathbf{x}; \quad W_{2,1} = -\int_{\Re^3} \rho_2 < \mathbf{x}, \frac{\partial \phi_1}{\partial \mathbf{x}} > d^3\mathbf{x}.$$

$$(7.18d)$$

A *wrong* application of the SVT to multi–component systems is to assume that $\ddot{I}_1/2 = 2K_1 + U_1 + U_{1,2}$, i.e., that in virialized multi–component systems $2K_1 = -U_1 - U_{1,2}$ instead of $2K_1 = -U_1 - W_{1,2}$. The error becomes apparent if we proceed to add the previous relations for the two components, thus obtaining $\ddot{I}/2 = (\ddot{I}_1 + \ddot{I}_2)/2 = 2K_1 + 2K_2 + U_1 + U_2 + U_{1,2} + U_{2,1} = 2K + U + U_{1,2} = 2K + U + U_{2,1}$. This should be compared with the correct eq. (7.18b). In turn, it is trivial to prove that

$$W_{1,2} + W_{2,1} = U_{2,1} = U_{1,2}, \qquad (7.18e)$$

and so, after summation of the two interaction energies, eq. (7.18a) is recovered.

8. PROJECTED DYNAMICS

When observed as astronomical objects, galaxies (and all their dynamical properties) appear projected on the plane of the sky. As a consequence, it is important to set the framework for a correct understanding of the relation between intrinsic dynamics (i.e., 3-dimensional; discussed in the previous chapters) and projected properties. This Chapter provides the reader with the basic tools needed for projecting the most important properties of stellar systems on a projection plane. In particular, we first describe the projection operator in phase–space, and then we derive the projected velocity moments associated with the 3-dimensional velocity moments discussed in Chapter 7. Moreover, the concepts of velocity profiles and line profiles are also presented. The derivation of the Projected Virial theorem (PVT), together with its application to the special case of spherical stellar systems, can be considered to be the central result of this Chapter.

8.1 The projection operator

Let us consider a given component of a multi–component collisionless stellar system, described in an inertial orthogonal reference system $S_0 = (O; \mathbf{e}_1, \mathbf{e}_2, \mathbf{e}_3)$ by its density $\rho = \rho(\mathbf{x}; t)$. In S_0 we introduce the orthogonal (time–independent) observer's system $S' = (O'; \mathbf{f}_1, \mathbf{f}_2, \mathbf{f}_3)$, and we assume that $O \equiv O'$ at all times. Throughout the Chapter, the repeated indices sum convention is used, so that $\mathbf{x} = x_i \mathbf{e}_i = x_1 \mathbf{e}_1 + x_2 \mathbf{e}_2 + x_3 \mathbf{e}_3$ is a vector in S_0 and $\boldsymbol{\xi} = \xi_j \mathbf{f}_j$ is the same vector in S'. We do not use the standard Euler angles[8.1] to specify the orientation of S' with respect to S_0 but instead we apply a 3-2-3 rotation. In this way ϑ and φ coincide with the polar coordinates of $\mathbf{f}_3$ in S_0 (being $\mathbf{e}_3$ the polar axis and $\mathbf{e}_1$ the azimuthal one):

$$\begin{cases} \boldsymbol{\xi} = \mathcal{R}\mathbf{x} \\ \boldsymbol{\nu} = \mathcal{R}\mathbf{v} \end{cases} \tag{8.1}$$

where $\boldsymbol{\nu}$ is the velocity as seen from S', and the rotation matrix is given by $\mathcal{R} = \mathcal{R}_3(\psi)\mathcal{R}_2(\vartheta)\mathcal{R}_3(\varphi)$. As a consequence of our choice, the unit vector $\mathbf{n} = \mathbf{f}_3$ from O to the observer is (see Fig. 8.1)

$$\mathbf{n}^{\mathrm{T}} = (\cos\varphi\sin\vartheta, \ \sin\varphi\sin\vartheta, \ \cos\vartheta), \tag{8.2}$$

where "T" means *transpose*, $0 \le \vartheta < \pi$ and $0 \le \varphi < 2\pi$. In this way $\xi_3 = <\mathbf{n}, \boldsymbol{\xi}>$. The definitions introduced so far are all *geometrical*, i.e. they do not involve the dynamics of the system. All the *dynamical* properties of the component ρ are known if its distribution function $f = f(\mathbf{x}, \mathbf{v}; t)$ is known. This function obeys the CBE, given by eqs. (6.1abc). In S', the DF is given by

$$f'(\boldsymbol{\xi}, \boldsymbol{\nu}; t) = f(\mathcal{R}^{\mathrm{T}}\mathbf{x}, \mathcal{R}^{\mathrm{T}}\mathbf{v}; t). \tag{8.3}$$

As discussed in Chapter 7, every *macroscopic* function $\overline{F}$ is defined by integration on the velocity space of its *microscopic* counterpart F, using f as weight function, and the resulting differential equation for $\overline{F}$ is given in eq. (7.3a). In the observer's reference frame S' the microscopic function F' is given by $F'(\boldsymbol{\xi}, \boldsymbol{\nu}; t) = F(\mathcal{R}^{\mathrm{T}}\mathbf{x}, \mathcal{R}^{\mathrm{T}}\mathbf{v}; t)$. The interested reader can prove, as a useful

---

[8.1] Starting from two coincident systems, the Euler angles are defined by a 3-1-3 rotation, i.e. by a rotation around $\mathbf{e}_3$ (φ), then around $\mathbf{e}_1'$ (ϑ) and finally around $\mathbf{e}_3''$ (ψ). As a result, the angles φ and ϑ are not the usual spherical angular coordinates of $\mathbf{e}_3''$ in S_0. See, e.g., [1.2], [1.5].

exercise, that $\overline{F}'(\xi;t) = \overline{F}(\mathcal{R}^T\mathbf{x};t)$, and $\overline{F'} = \int f'F'd^3\nu = (\int fFd^3\mathbf{v})' = \overline{F}'$, where the last identity is obtained from the orthogonality of $\mathcal{R}$.

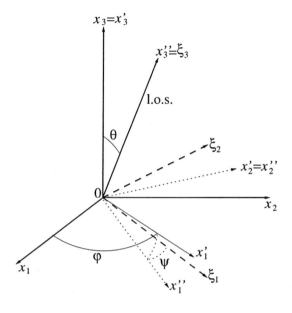

Figure 8.1

The *projection* of a given property $\overline{F} = \overline{F}(\mathbf{x};t)$, associated with our system via eq. (7.1a), onto the *projection plane* $\pi' = (\xi_1, \xi_2)\perp\mathbf{n}$, can be formally introduced by the following

Definition 8.1 [Projection] *In a general system, we define*

$$\Sigma(\xi_1, \xi_2; t)\overline{F}_{\text{los}}(\xi_1, \xi_2; t) := \int_{-\infty}^{\infty} \rho'(\xi;t)\overline{F}'(\xi;t)d\xi_3, \qquad (8.4a)$$

where

$$\Sigma(\xi_1, \xi_2; t) := \int_{-\infty}^{\infty} \rho'(\xi;t)d\xi_3. \qquad (8.4b)$$

The subscript "los" means line–of–sight *(hereafter, l.o.s.).*

Note that the projection operator has been defined here for macroscopic functions $\overline{F}$: in fact, the operations of projection and average over velocity space do not commute. The following general result holds:

Theorem 8.2 *For any* $\overline{F}$

$$\int_{\pi'} \Sigma(\xi_1, \xi_2; t)\overline{F}_{\text{los}}(\xi_1, \xi_2; t)d\xi_1 d\xi_2 = \int_{\Re^3} \rho(\mathbf{x}; t)\overline{F}(\mathbf{x}; t)d^3\mathbf{x}. \qquad (8.5)$$

PROOF   The result is obtained by inserting eq. (8.4a) in eq. (8.5), changing variables according to eq. (8.1), and using the orthogonality of $\mathcal{R}$.   ◁

Note that, by assuming $\overline{F} = 1$, one obtains that the surface integral of the projected density equals the total mass associated with the density distribution ρ, *independently* of the particular orientation of the observer.

8.2 Projected velocity moments

In Chapter 7 was shown that a special status, within the infinite family of microscopic functions F that can be constructed, is assumed by the velocity monomials of any order. In fact, these functions enter directly in the Virial theorem, one of the basic results of Dynamics. Here we take F to be the N–th power of the $\mathbf{x}$ component along $\mathbf{n}$: the results presented in this Section are both interesting "per se", linking the intrinsic velocity moments with the projected ("observables") ones, and also being a step to the derivation of the Projected Virial theorem (derived in Section 8.4).

Definition 8.3 [Projected velocity field] *Let*

$$F_1(\mathbf{v}) :=< \mathbf{n}, \mathbf{v} >^N = (n_i v_i)^N := v_{\text{p}}^N, \qquad (8.6)$$

where $N \geq 0$ is an integer. With $\overline{v_{\text{p}}^N}$ we indicate the mean over velocities in phase–space of v_{p}^N, and we call this quantity the N–th order projected velocity field *along* $\mathbf{n}$ *at position* $\mathbf{x}$; *its projection according to eq. (8.4a),* $\overline{v_{\text{p}}^N}_{\text{los}}$, *is the l.o.s. projected velocity field of order N.*

The following results are of immediate proof:

Theorem 8.3a *For $N \geq 0$,*

$$v_{\text{p}}^N = \sum_{K=0}^{N} \sum_{L=0}^{K} \binom{N}{K}\binom{K}{L}(n_3^{N-K}n_2^{K-L}n_1^L)(v_3^{N-K}v_2^{K-L}v_1^L), \qquad (8.7a)$$

$$\overline{v_{\text{p}}^N} = \sum_{K=0}^{N} \sum_{L=0}^{K} \binom{N}{K}\binom{K}{L}(n_3^{N-K}n_2^{K-L}n_1^L)\overline{(v_3^{N-K}v_2^{K-L}v_1^L)}, \qquad (8.7b)$$

$$\overline{v_{\text{p}}^{N}}_{\text{los}} = \sum_{K=0}^{N} \sum_{L=0}^{K} \binom{N}{K}\binom{K}{L} (n_3^{N-K} n_2^{K-L} n_1^{L}) \overline{(v_3^{N-K} v_2^{K-L} v_1^{L})}_{\text{los}}, \qquad (8.7c)$$

where $\binom{a}{b}$ are the standard binomial coefficients.

For example, for $N = 1$

$$\overline{v_{\text{p}}}(\mathbf{x}; t) = \overline{< \mathbf{n}, \mathbf{v} >} = n_i \overline{v_i}, \qquad (8.8)$$

because for $N = 1$ the projection of $\mathbf{v}$ over $\mathbf{n}$ and the integration over velocity space commute; for a given observer orientation $\overline{v_{\text{p}}}$ (as a function of $\mathbf{x}$, and according to Definition 7.3) is called the *projected streaming velocity field*. A positive $\overline{v_{\text{p}}}$ means that the component of the projected streaming velocity along $\mathbf{n}$ points toward the observer. Finally, the projection of $\overline{v_{\text{p}}}$ according to eq. (8.4a), $\overline{v_{\text{p}}}_{\text{los}}$, is the so–called *l.o.s. streaming velocity field*. Note that, if the system does not have intrinsic streaming motions, i.e. $\overline{v_i} = 0$ for $i = 1, 2, 3$, then $\overline{v_{\text{p}}} = \overline{v_{\text{p}}}_{\text{los}} = 0$, but the converse is not true. For a given orientation $\overline{v_{\text{p}}}$ and $\overline{v_{\text{p}}}_{\text{los}}$ may vanish, but the system can still possess streaming motions. Moreover, under particular circumstances, $\overline{v_{\text{p}}}_{\text{los}}$ can vanish even if $\overline{v_{\text{p}}} \neq 0$.

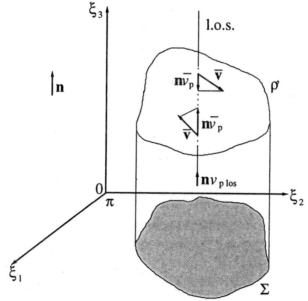

Figure 8.2

A second function over the phase–space, analogous to F_1 in eq. (8.6), is naturally introduced:

Definition 8.4 [Projected velocity dispersion field] *Let*

$$F_2(\mathbf{v}) := (v_{\mathrm{p}} - \overline{v_{\mathrm{p}}})^N := \sigma_{\mathrm{p}}^N, \qquad (8.9)$$

where $N \geq 0$ is an integer. With $\overline{\sigma_{\mathrm{p}}^N}$ we indicate the mean over velocities in phase–space of σ_{p}^N, and we call this quantity the N–th order projected velocity dispersion field *along $\mathbf{n}$ at position $\mathbf{x}$; its projection according to eq. (8.4a), $\overline{\sigma_{\mathrm{p}}^N}_{\mathrm{los}}$, is the* l.o.s. projected velocity dispersion field *of order N.*

The following results are of immediate proof:

Theorem 8.4a *For $N \geq 0$,*

$$\sigma_{\mathrm{p}}^N = \sum_{K=0}^{N} \binom{N}{K} (-1)^K v_{\mathrm{p}}^{N-K} \overline{v_{\mathrm{p}}}^K, \qquad (8.10a)$$

$$\overline{\sigma_{\mathrm{p}}^N} = \sum_{K=0}^{N} \binom{N}{K} (-1)^K \overline{v_{\mathrm{p}}^{N-K}} \, \overline{v_{\mathrm{p}}}^K, \qquad (8.10b)$$

$$\overline{\sigma_{\mathrm{p}}^N}_{\mathrm{los}} = \sum_{K=0}^{N} \binom{N}{K} (-1)^K (\overline{v_{\mathrm{p}}^{N-K}} \, \overline{v_{\mathrm{p}}}^K)_{\mathrm{los}}. \qquad (8.10c)$$

From the above identities it follows that, for any N, σ_{p}^N and $\overline{\sigma_{\mathrm{p}}^N}$ can be expressed as simple functions of v_{p}^K and $\overline{v_{\mathrm{p}}^K}$ with $K = 0, 1, ..., N$. Note that, if $\overline{v_{\mathrm{p}}} = 0$, then $\sigma_{\mathrm{p}}^N = v_{\mathrm{p}}^N$ and so $\overline{\sigma_{\mathrm{p}}^N}_{\mathrm{los}} = \overline{v_{\mathrm{p}}^N}_{\mathrm{los}}$. Moreover, note that for $N = 1$ $\overline{\sigma_{\mathrm{p}}} = \overline{\sigma_{\mathrm{p}los}} = 0$ always. A particularly important case is obtained for $N = 2$, for which

$$\overline{\sigma_{\mathrm{p}}^2}(\mathbf{x}; t) = n_i n_j \sigma_{ij}^2, \qquad (8.11)$$

where σ_{ij}^2 is the velocity dispersion tensor.

When observing a given system, the observed l.o.s. velocity dispersion field (see next Definition) does not coincide in general with $\overline{\sigma_{\mathrm{p}}^N}_{\mathrm{los}}$. In fact, the observations give the l.o.s. velocity dispersion with respect to the observed l.o.s. streaming velocity field $\overline{v_{\mathrm{p}los}}$, *not* the l.o.s. projected velocity dispersion field $\overline{\sigma_{\mathrm{p}}^N}_{\mathrm{los}}$, which is defined by using $\overline{v_{\mathrm{p}}}$ instead of $\overline{v_{\mathrm{p}los}}$. This leads us to introduce a third and last function over the phase space, analogous to F_1 and F_2:

Definition 8.5 [Local l.o.s. velocity dispersion field] *Let*

$$F_3(\mathbf{v}) = (v_p - \overline{v_{p\,los}})^N = \sigma_1^N, \tag{8.12}$$

where $N \geq 0$ is an integer. With $\overline{\sigma_1^N}$ we indicate the mean over velocities in phase–space of σ_1^N, and we call this quantity the N–th order local l.o.s. velocity dispersion field along $\mathbf{n}$ at position $\mathbf{x}$; its projection according to eq. (8.4a), $\overline{\sigma_1^N}_{los}$, is the l.o.s. velocity dispersion field of order N.

The following results are of immediate proof:

Theorem 8.5a *For $N \geq 0$,*

$$\sigma_1^N = \sum_{K=0}^{N} \binom{N}{K}(-1)^K v_p^{N-K} \left(\overline{v_{p\,los}}\right)^K, \tag{8.13a}$$

$$\overline{\sigma_1^N} = \sum_{K=0}^{N} \binom{N}{K}(-1)^K \overline{v_p^{N-K}} \left(\overline{v_{p\,los}}\right)^K, \tag{8.13b}$$

$$\overline{\sigma_1^N}_{los} = \sum_{K=0}^{N} \binom{N}{K}(-1)^K \left(\overline{v_p^{N-K}}\right)_{los} \left(\overline{v_{p\,los}}\right)^K. \tag{8.13c}$$

From the above identities it follows that σ_1^N and $\overline{\sigma_1^N}$ can be expressed for any N as simple functions of v_p^K, $\overline{v_p^K}$, and $\overline{v_{p\,los}}$ with $K = 0, 1, ..., N$. Moreover, $\overline{\sigma_1^N}_{los}$ can be expressed as a simple function of the family of $\overline{v_p^K}_{los}$ with $K = 0, 1, ..., N$. Note that, if the system is characterized by $\overline{v_{p\,los}} = 0$, then $\sigma_1^N = v_p^N$ and so $\overline{\sigma_1^N}_{los} = \overline{v_p^N}_{los}$. Moreover, if $\overline{v_p} = 0$, then $\sigma_1^N = v_p^N = \sigma_p^N$ and so $\overline{\sigma_1^N}_{los} = \overline{v_p^N}_{los} = \overline{\sigma_p^N}_{los}$. In particular, the last identity above holds for a system with a vanishing streaming velocity field. Finally, note that for $N = 1$, $\overline{\sigma_{1\,los}} = 0$ always.

As shown, the fields $\overline{\sigma_p^N}$ and $\overline{\sigma_1^N}$ (and the associated l.o.s. fields) can be expressed as functions of the fields $\overline{v_p^K}$ with $K = 0, 1, ..., N$ (and their l.o.s. projections). Is the opposite true? This is a very important question, because from an observational point of view we can assume to have access to the members of the family σ_1^N only, together with $\overline{v_{\mathrm{Plos}}}$. The following Theorem provides the answer.

Theorem 8.6 *For $N \geq 0$,*

$$v_{\mathrm{p}}^N = \sum_{K=0}^{N} \binom{N}{K} \overline{v_{\mathrm{p}}}^{N-K} \sigma_{\mathrm{p}}^K = \sum_{K=0}^{N} \binom{N}{K} (\overline{v_{\mathrm{Plos}}})^{N-K} \sigma_1^K, \qquad (8.14a)$$

$$\overline{v_{\mathrm{p}}^N} = \sum_{K=0}^{N} \binom{N}{K} \overline{v_{\mathrm{p}}}^{N-K} \overline{\sigma_{\mathrm{p}}^K} = \sum_{K=0}^{N} \binom{N}{K} (\overline{v_{\mathrm{Plos}}})^{N-K} \overline{\sigma_1^K}, \qquad (8.14b)$$

$$\overline{v_{\mathrm{p}}^N}_{\mathrm{los}} = \sum_{K=0}^{N} \binom{N}{K} (\overline{v_{\mathrm{p}}}^{N-K} \overline{\sigma_{\mathrm{p}}^K})_{\mathrm{los}} = \sum_{K=0}^{N} \binom{N}{K} (\overline{v_{\mathrm{Plos}}})^{N-K} \overline{\sigma_1^K}_{\mathrm{los}}. \qquad (8.14c)$$

$$\sigma_{\mathrm{p}}^N = \sum_{K=0}^{N} \binom{N}{K} (-1)^K \sigma_1^{N-K} \overline{\sigma_1}^K, \qquad (8.15a)$$

$$\overline{\sigma_{\mathrm{p}}^N} = \sum_{K=0}^{N} \binom{N}{K} (-1)^K \overline{\sigma_1^{N-K}} \overline{\sigma_1}^K, \qquad (8.15b)$$

$$\overline{\sigma_{\mathrm{p}}^N}_{\mathrm{los}} = \sum_{K=0}^{N} \binom{N}{K} (-1)^K (\overline{\sigma_1^{N-K}} \overline{\sigma_1}^K)_{\mathrm{los}}. \qquad (8.15c)$$

$$\sigma_1^N = \sum_{K=0}^{N} \binom{N}{K} \sigma_{\mathrm{p}}^{N-K} \overline{\sigma_1}^K, \qquad (8.16a)$$

$$\overline{\sigma_1^N} = \sum_{K=0}^{N} \binom{N}{K} \overline{\sigma_{\mathrm{p}}^{N-K}} \overline{\sigma_1}^K, \qquad (8.16b)$$

$$\overline{\sigma_1^N}_{\mathrm{los}} = \sum_{K=0}^{N} \binom{N}{K} (\overline{\sigma_{\mathrm{p}}^{N-K}} \overline{\sigma_1}^K)_{\mathrm{los}}. \qquad (8.16c)$$

PROOF   The first set of identities in eqs. (8.14abc) is obtained by rewriting eq. (8.6) as $F_1(\mathbf{v}) = (v_{\mathrm{p}} - \overline{v_{\mathrm{p}}} + \overline{v_{\mathrm{p}}})^N$, and by using eq. (8.9). The second set of identities in eqs. (8.14abc) is obtained by rewriting eq. (8.6) as $F_1(\mathbf{v}) = (v_{\mathrm{p}} - \overline{v_{\mathrm{Plos}}} + \overline{v_{\mathrm{Plos}}})^N$, and by using eq. (8.12). The set of identities in eqs. (8.15abc) is obtained by rewriting eq. (8.9) as $F_2(\mathbf{v}) = (v_{\mathrm{p}} - \overline{v_{\mathrm{Plos}}} + \overline{v_{\mathrm{Plos}}} - \overline{v_{\mathrm{p}}})^N$, and by using eq. (8.12). Finally, the set of identities in eqs. (8.16abc) is obtained by rewriting eq. (8.12) as $F_3(\mathbf{v}) = (v_{\mathrm{p}} - \overline{v_{\mathrm{p}}} + \overline{v_{\mathrm{p}}} - \overline{v_{\mathrm{Plos}}})^N$, and by using eq. (8.9).   ◁

Note that from eq. (8.14c) it follows that $\overline{v_{\mathrm{p}}^N}_{\mathrm{los}}$ can be expressed as a simple function of $\overline{v_{\mathrm{p}}}_{\mathrm{los}}$ and the fields $\overline{\sigma_{\mathrm{p}}^K}_{\mathrm{los}}$ with $K = 0, 1, ..., N$. Moreover, from eq. (8.14b) with $N = 2$,

$$\overline{v_{\mathrm{p}}^2} = \overline{v_{\mathrm{p}}}^2 + \overline{\sigma_{\mathrm{p}}^2}, \tag{8.17a}$$

and, from eq. (8.14c),

$$\overline{\sigma_{\mathrm{l}}^2}_{\mathrm{los}} = \overline{v_{\mathrm{p}\,\mathrm{los}}^2} - (\overline{v_{\mathrm{p}\,\mathrm{los}}})^2 = (\overline{v_{\mathrm{p}}^2})_{\mathrm{los}} + \overline{\sigma_{\mathrm{p}\,\mathrm{los}}^2} - (\overline{v_{\mathrm{p}\,\mathrm{los}}})^2, \tag{8.17b}$$

where the last identity follows from the l.o.s. projection of eq. (8.17a). Note that, even in a system with no intrinsic velocity dispersion (i.e., $\overline{\sigma_{\mathrm{p}\,\mathrm{los}}^2} = 0$), $\overline{\sigma_{\mathrm{l}}^2}_{\mathrm{los}}$ can be greater than zero. A stellar disk where all stars follow circular orbits around its center (a "cold" disk) in the same sense, when projected on a plane perpendicular to the disk itself, is the simplest example of this situation.

8.3 Velocity profiles

A natural question arises: how are the fields $\overline{v_{\mathrm{p}}}_{\mathrm{los}}$ and $\overline{\sigma_{\mathrm{l}}^N}_{\mathrm{los}}$ related to the distribution function f' referred to the S' frame of reference? In other words, what is the physical meaning of the l.o.s. velocity dispersion and streaming fields? We start the discussion by introducing the concept of Velocity Profile (VP), for which a physical interpretation follows immediately from its

Definition 8.7 [Velocity profile] *The* velocity profile (VP) *at a point in the projection plane* $(\xi_1, \xi_2) \in \pi'$ *is defined as*

$$\Sigma(\xi_1, \xi_2; t) \mathrm{VP}(\xi_1, \xi_2, v_{\mathrm{p}}; t) = \int f'(\boldsymbol{\xi}, \boldsymbol{\nu}; t)\, d\nu_1 d\nu_2 d\xi_3. \tag{8.18}$$

Note that by definition the units of VP are those of an inverse velocity. Moreover, $\int \mathrm{VP}\, dv_{\mathrm{p}} = \int \rho'\, d\xi_3 / \Sigma = 1$, i.e., the integral of VP over all possible velocities is normalized to unity. Needles to say, VP vanishes for sufficiently high values of $|v_{\mathrm{p}}|$. We are now in the position to answer the question posed above:

Theorem 8.8 *The following identities hold:*

$$\overline{v_{\mathrm{p}}}_{\mathrm{los}}(\xi_1, \xi_2; t) = \int_{\Re} v_{\mathrm{p}} \mathrm{VP} dv_{\mathrm{p}}, \tag{8.19a}$$

$$\overline{\sigma_{1}^{N}}_{\text{los}}(\xi_1, \xi_2; t) = \int_{\Re} (v_{\text{p}} - \overline{v_{\text{p}}}_{\text{los}})^N \, \text{VP} dv_{\text{p}}. \qquad (8.19b)$$

PROOF   From eqs. (8.18) and (8.6) it follows that

$$\Sigma \int v_{\text{p}} \, \text{VP} dv_{\text{p}} = \int v_{\text{p}} f' d^3 \nu d\xi_3 = \int (\rho \overline{v_{\text{p}}})' \, d\xi_3 = \Sigma \overline{v_{\text{p}}}_{\text{los}}.$$

From eqs. (8.18) and (8.12), it follows that

$$\Sigma \int (v_{\text{p}} - \overline{v_{\text{p}}}_{\text{los}})^N \, \text{VP} dv_{\text{p}} = \int (v_{\text{p}} - \overline{v_{\text{p}}}_{\text{los}})^N f' d^3 \nu d\xi_3 = \int (\rho \overline{\sigma_1^N})' \, d\xi_3 = \Sigma \overline{\sigma_1^N}_{\text{los}}.$$

◁

We now move another step toward the observational world, by linking the velocity profile with a function available (in principle!) to astronomers dealing with stellar spectra, i.e., the so–called *line profile* (LP). Let us assume for simplicity that all stars associated with the distribution function f have the same *mass–to–light* ratio Υ, and that, when observed at rest, a given spectral line is described by the (normalized) function $P_0 = P_0(\lambda)$, with $\int_0^\infty P_0(\lambda) d\lambda = 1$. In the resulting LP, each star contributes with P_0 red or blue shifted according to the sign of v_{p}. As a result of the assumed orientation of S', a negative v_{p} corresponds to a red–shift. For $|v_{\text{p}}| \ll c$ (where c is the speed of light), $\lambda' = \lambda(1 - v_{\text{p}}/c)$, and so the (normalized) shifted spectral line profile is given by $P_v(\lambda) = P_0[\lambda/(1 - v_{\text{p}}/c)](1 - v_{\text{p}}/c)^{-1}$. As a consequence,

Definition 8.9 [Line profile] *The* line–profile (LP) *is defined as*[8.2]

$$\text{LP}(\xi_1, \xi_2, \lambda; t) = \int_{\Re} P_0 \left(\frac{\lambda}{1 - v_{\text{p}}/c} \right) \frac{\text{VP} \, dv_{\text{p}}}{1 - v_{\text{p}}/c}. \qquad (8.20a)$$

Because of the normalization condition on VP and P_0, then the line profile also is normalized to unity, $\int_0^\infty \text{LP} \, d\lambda = 1$. As a simple example of line

---

[8.2]   Note that the definition is well–posed. In fact, by summing the contribution of individual stars

$$\Sigma \text{LP} = \int d\xi_3 \int f' \, d\nu_1 d\nu_2 \, P_v \, dv_{\text{p}}$$

$$= \int P_v \, dv_{\text{p}} \int f' \, d\nu_1 d\nu_2 d\xi_3 = \Sigma \int P_v \, \text{VP} \, dv_{\text{p}},$$

and so eq. (8.20a) can be recovered starting from the distribution function.

profile, let us assume $f' = f'(\xi, \nu_1, \nu_2; t)\delta(\nu_3 - v_{\rm p}^0)$, i.e., all stars are moving with the same velocity along the l.o.s.. In this case, VP $= \delta(\nu_3 - v_{\rm p}^0)$, and LP $= P_0[\lambda/(1 - v_{\rm p}^0/c)](1 - v_{\rm p}^0/c)^{-1}$. As a second example, let us assume $P_0 = \delta(\lambda - \lambda_0)$, and so $P_v = A\delta[\lambda/(1 - v_{\rm p}/c) - \lambda_0](1 - v_{\rm p}/c)^{-1}$. The only contribution to the integral in eq. (8.20a) comes from $v_{\rm p} = (1 - \lambda/\lambda_0)c$, and so simple calculations show that LP $= (c\lambda_0){\rm VP}[\xi_1, \xi_2, (1 - \lambda/\lambda_0)c; t]$, where we used the standard rule for the change of variable in the Dirac distribution δ. The normalization property can be easily verified also in this case.

Let now $\lambda_0 = \int_0^\infty P_0(\lambda)\lambda d\lambda$ be the characteristic wavelength (in the laboratory rest–frame) associated with the P_0 profile, and $\sigma_0^2 = \int_0^\infty P_0(\lambda)(\lambda - \lambda_0)^2 d\lambda$ its dispersion. The same quantities can be defined using the "observed" line profiles, instead of P_0, and the reader can show that

$$\lambda_v := \int_0^\infty {\rm LP}(\lambda)\lambda d\lambda = \left(1 - \frac{\overline{v_{\rm p}}_{\rm los}}{c}\right)\lambda_0, \qquad (8.20b)$$

$$\sigma_v^2 := \int_0^\infty {\rm LP}(\lambda)(\lambda - \lambda_v)^2 d\lambda = \left(1 - \frac{\overline{v_{\rm p}}_{\rm los}}{c}\right)^2 \sigma_0^2 + \frac{(\lambda_0^2 + \sigma_0^2)\overline{\sigma_1^2}_{\rm los}}{c^2}. \qquad (8.20c)$$

Similar relations between higher order moments of the line profiles and l.o.s. velocity profiles (streaming and dispersion) can be derived in the same way. Thus, the line profiles represent the natural connection between the observational astronomy and the projected dynamical properties of stellar systems.

8.4 The Projected Virial theorem (PVT)

In order to derive and discuss the Projected Virial theorem (PVT), we need to determine the evolution equation for $\overline{v_{\rm p}^N}$. In fact, from eqs. (8.6) and (7.3a), it is immediate to derive the following

Theorem 8.10 *The equation satisfied by $\overline{v_{\rm p}^N}$ is*

$$\frac{\partial \overline{\rho v_{\rm p}^N}}{\partial t} + \frac{\partial \overline{\rho v_{\rm p}^N v_i}}{\partial x_i} = -N n_i \overline{\rho v_{\rm p}^{N-1}\frac{\partial \phi_{\rm T}}{\partial x_i}}. \qquad (8.21)$$

We now generalize the definition of total kinetic energy as follows:

Definition 8.11 [L.o.s. N–th order total kinetic energy] *The l.o.s. N–th order total kinetic energy is defined, for $N \geq 1$, as*

$$K_{\rm los}^{(N)}(t) := \frac{1}{N}\int \overline{\Sigma v_{\rm p}^N}_{\rm los} d\xi_1 d\xi_2. \qquad (8.22a)$$

The following result holds:

Theorem 8.12

$$K_{\text{los}}^{(N)}(t) = \frac{1}{N} \int \rho \overline{v_{\text{p}}^N} d^3\mathbf{x}, \tag{8.22b}$$

$$K_{\text{los}}^{(1)}(t) = n_i P_i, \tag{8.22c}$$

$$K_{\text{los}}^{(2)}(t) = n_i n_j K_{ij}. \tag{8.22d}$$

where P_i is the i-th ($i = 1, 2, 3$) component of the linear momentum associated with ρ.

PROOF   Equations (8.22c) and (8.22d) are a direct consequence of eqs. (8.22b) and (8.7b), for $N = 1$ and $N = 2$ respectively. Equation (8.22b) is a direct consequence of eqs. (8.22a) and (8.5).   ◁

After multiplication of eq. (8.21) by x_j ($j = 1, 2, 3$) and integration over the configuration space, one obtains three equations. Multiplying each of these equations for n_j, summing, and using eq. (8.22b), we finally obtain the N–th order PVT:

Theorem 8.13 [PVT] *For $N \geq 1$,*

$$\frac{dI_{\text{los}}^{(N)}}{dt} = (N+1)K_{\text{los}}^{(N+1)} + NW_{\text{T,los}}^{(N-1)} \tag{8.23a}$$

where

$$W_{\text{T,los}}^{(N-1)} := n_i n_j W_{\text{T},ij}^{(N-1)}, \quad W_{\text{T},ij}^{(N-1)} := -\int_{\Re^3} \rho \overline{v_{\text{p}}^{N-1}} x_j \frac{\partial \phi_{\text{T}}}{\partial x_i} d^3\mathbf{x} \tag{8.23b}$$

$$I_{\text{los}}^{(N)} = \int_{\Re^3} <\mathbf{n}, \mathbf{x}> \rho \overline{v_{\text{p}}^N} d^3\mathbf{x} \tag{8.23c}$$

The projection of the usual (second order) Virial theorem (eqs. [7.17ab]) is then immediately obtained by setting $N = 1$ in eq. (8.23a).

Theorem 8.14 [Second order PVT] *The second order PVT reads:*

$$\frac{n_i n_j \ddot{I}_{ij}}{2} = 2K_{\text{los}}^{(2)} + n_i n_j \left(U_{ij} + \frac{W_{ij} + W_{ji}}{2} \right), \tag{8.24}$$

where U_{ij} and W_{ij} are given by eqs. (7.14ab).

PROOF   From eq. (7.16),

$$\dot{I}_{ij}n_in_j = \int \rho n_i n_j (x_i\overline{v_j} + x_j\overline{v_i})d^3\mathbf{x} = 2I_{\text{los}}^{(1)}.$$

Moreover, $W_{\text{T},ij}^{(N-1)}$ do not depend on velocity, and from eqs. (7.14ab), $W_{\text{T},ij}^{(0)} = U_{ij}+W_{ij}$.
By exchanging i with j, and by using the symmetry of U_{ij}, and I_{ij}, the result is proved.
◁

8.5 The PVT for spherical systems

When applying the PVT to a system with no special symmetries the direction of the observer line–of–sight must be specified. For example, one could use eq. (8.24) to derive the value of the l.o.s. kinetic energy, without solving the Jeans equations, from the potential energy tensors and the three components of **n**. In contrast, in the highly idealized (and thus frequently used) situation of spherical symmetry, the PVT *cannot* depend on the observation direction. We now discuss this case by calculating the *angular mean* over the solid angle of the PVT. Of course, in spherically symmetric systems the value of the angular mean equals the l.o.s. value. A different approach to the solution of this problem will be presented (under more restrictive hypotheses) in Chapter 11 (Theorem 11.8).

Let us consider a fictitious "observer" that can move all around a stellar system, observing some projected property $\overline{F}_{\text{los}}$. In general, this property will depend on the l.o.s. direction **n**. The angular mean of $\overline{F}_{\text{los}}$ over the solid angle is defined simply as:

Definition 8.15 [Angular mean] *For a given $\overline{F}_{\text{los}}$, its associated* angular mean *is defined as*

$$[\overline{F}_{\text{los}}]_\Omega := \frac{1}{4\pi}\int_{4\pi} \overline{F}_{\text{los}}d^2\Omega = \frac{1}{4\pi}\int_0^\pi \int_0^{2\pi} \overline{F}_{\text{los}} \sin\vartheta\, d\vartheta d\varphi; \qquad (8.25)$$

the last expression is determined by our choice of the orientation of S'.

Our task is to obtain the angular mean of the N–th order PVT, as given in eq. (8.23a). The following result will be useful in later calculations:

Lemma 8.16 *For $N \geq 1$,*

$$\overline{[v_{\rm p}^{N+1}]}_\Omega = \overline{[v_{\rm p}^{N+1}]}_\Omega, \tag{8.26a}$$

$$\overline{[n_i v_{\rm p}^{N}]}_\Omega = \frac{1}{N+1} \frac{\partial \overline{[v_{\rm p}^{N+1}]}_\Omega}{\partial v_i}, \tag{8.26b}$$

$$\overline{[n_i n_j v_{\rm p}^{N-1}]}_\Omega = \frac{1}{N(N+1)} \frac{\partial^2 \overline{[v_{\rm p}^{N+1}]}_\Omega}{\partial v_i \partial v_j}. \tag{8.26c}$$

PROOF   The identities above are a direct consequence of eqs. (8.6) and (8.7a).   ◁

From Lemma 8.16 and from eqs. (8.23abc), the determination of the angular mean of the PVT of N–th order requires only the calculation of $[v_{\rm p}^N]_\Omega$ for a generic N. From eqs. (8.7abc) we need to integrate products of trigonometric functions with integer exponents. The algebra involved is rather lengthy, so some preliminaries are necessary.

Lemma 8.16 *For a and b non–negative integers, let*
$I_\phi(a,b) := \int_0^\phi \sin^a x \cos^b x \, dx$. *The following identities hold:*

$$I_{\pi/2}(a,b) = \frac{1}{2}{\rm B}\left(\frac{a+1}{2}, \frac{b+1}{2}\right) = I_{\pi/2}(b,a), \tag{8.27a}$$

$$I_\pi(a,b) = [1+(-1)^b]I_{\pi/2}(a,b), \tag{8.27b}$$

$$I_{2\pi}(a,b) = [1+(-1)^{a+b}][1+(-1)^b]I_{\pi/2}(a,b), \tag{8.27c}$$

where ${\rm B}(x,y)$ is the complete Euler Beta function.

PROOF   Equation (8.27a) is obtained from eq. (8.380.2) of [4.7]. Equation (8.27b) is obtained by reducing the trigonometric functions in the range $(\pi/2, \pi)$ to the first quadrant. Equation (8.27c) is obtained by reducing the trigonometric functions in the range $(\pi, 2\pi)$ to the upper half–plane and by using eq. (8.27b).   ◁

We calculate now the angular mean of a generic angular coefficient in eqs. (8.7abc), using eq. (8.25).

Lemma 8.18 *For $N \geq K \geq L \geq 0$,*

$$[n_3^{N-K} n_2^{K-L} n_1^L]_\Omega = \frac{[1+(-1)^{N-K}][1+(-1)^K][1+(-1)^L]}{8\pi(N+1)} \times$$
$$\frac{\Gamma\left(\frac{N-K+1}{2}\right)\Gamma\left(\frac{K-L+1}{2}\right)\Gamma\left(\frac{L+1}{2}\right)}{\Gamma\left(\frac{N+1}{2}\right)}, \tag{8.28}$$

where $\Gamma(x)$ is the complete Euler Gamma function.

PROOF   For $N \geq K \geq L \geq 0$, and from eq. (8.2),

$$4\pi[n_3^{N-K}n_2^{K-L}n_1^L]_\Omega = I_\pi(K+1, N-K)I_{2\pi}(K-L, L).$$

After the reduction to the first quadrant by using eqs. (8.27bc), one obtains

$$I_{\pi/2}(K+1, N-K)I_{\pi/2}(K-L, L) = \frac{1}{4}B\left(\frac{K+2}{2}, \frac{N-K+1}{2}\right) \times$$
$$B\left(\frac{K-L+1}{2}, \frac{L+1}{2}\right).$$

From the transformation rule $B(x, y) = \Gamma(x)\Gamma(y)/\Gamma(x+y)$ and $\Gamma(x+1) = x\Gamma(x)$, the result is proved.   ◁

A further simplification is obtained by considering the binomial coefficients in eqs. (8.7abc).

Lemma 8.19 *For $N \geq K \geq L \geq 0$*

$$\binom{N}{K}\binom{K}{L}[n_3^{N-K}n_2^{K-L}n_1^L]_\Omega = \frac{[1+(-1)^{N-K}][1+(-1)^K][1+(-1)^L]}{8(N+1)} \times$$
$$\frac{\Gamma\left(\dfrac{N+2}{2}\right)}{\Gamma\left(\dfrac{N-K+2}{2}\right)\Gamma\left(\dfrac{K-L+2}{2}\right)\Gamma\left(\dfrac{L+2}{2}\right)}.\quad(8.29)$$

PROOF   The binomial coefficient are written as

$$\binom{x}{y} = \frac{\Gamma(x+1)}{\Gamma(y+1)\Gamma(x-y+1)};$$

then the duplication formula for the Gamma function,

$$\frac{\Gamma(x)}{\Gamma(x/2)} = \frac{2^{x-1}}{\sqrt{\pi}}\Gamma\left(\frac{x+1}{2}\right)$$

is applied.   ◁

We are now ready to prove the following

Theorem 8.20 [Angular mean of v_p^N] *Let n be a non–negative integer. For $N = 2n + 1$,*

$$[v_p^N]_\Omega = 0.\quad(8.30a)$$

For $N = 2n$

$$[v_p^N]_\Omega = \frac{1}{2n+1} \sum_{k=0}^{n} \sum_{l=0}^{k} \binom{n}{k}\binom{k}{l} (v_3^{n-k} v_2^{k-l} v_1^l)^2. \qquad (8.30b)$$

PROOF   From eq. (8.29) only the terms with even K and L have angular mean different from zero. If $N = 2n + 1$ then $N - K$ is odd, and eq. (8.30a) is proved. By setting $N = 2n$, $K = 2k$, and $L = 2l$ in eq. (8.29), and by using again the relation between binomial coefficients and Gamma function, eq. (8.30b) is proved.   ◁

For example, for $n = 1$, $[v_p^2]_\Omega = (v_1^2 + v_2^2 + v_3^2)/3$, and so

$$[v_p^2]_\Omega = \frac{\overline{v_1}^2 + \overline{v_2}^2 + \overline{v_3}^2}{3} + \frac{\sigma_{11}^2 + \sigma_{22}^2 + \sigma_{33}^2}{3}. \qquad (8.31)$$

Finally, we can apply the angular mean to the N-th order PVT given in eqs. (8.23abc):

Theorem 8.21 [Angular mean of the N-th order PVT]
For $N \geq 1$, the angular mean of the PVT of order N is given by

$$\frac{d[I_{\text{los}}^{(N)}]_\Omega}{dt} = (N+1)[K_{\text{los}}^{(N+1)}]_\Omega + N[W_{\text{T,los}}^{(N-1)}]_\Omega, \qquad (8.32a)$$

where

$$[I_{\text{los}}^{(N)}]_\Omega = \frac{1}{N+1} \int_{\Re^3} \rho x_i \frac{\overline{\partial [v_p^{N+1}]_\Omega}}{\partial v_i} d^3\mathbf{x}, \qquad (8.32b)$$

$$[W_{\text{T,los}}^{(N-1)}]_\Omega = -\frac{1}{N(N+1)} \int_{\Re^3} \rho x_j \frac{\partial \phi_T}{\partial x_i} \frac{\overline{\partial^2 [v_p^{N+1}]_\Omega}}{\partial v_i \partial v_j} d^3\mathbf{x}, \qquad (8.32c)$$

$$[K_{\text{los}}^{(N+1)}]_\Omega = \frac{1}{N+1} \int_{\Re^3} \rho \overline{[v_p^{N+1}]_\Omega} d^3\mathbf{x}. \qquad (8.32d)$$

PROOF   Equation (8.32a) is the angular mean of eq. (8.23a). Equations (8.32bcd) are derived from eqs. (8.26abc).   ◁

For example, for $N = 1$,

$$[K_{\text{los}}^{(2)}]_\Omega = \frac{K}{3}, \qquad (8.33)$$

i.e., the angular mean of the total l.o.s. kinetic energy equals *one third* of the total kinetic energy. The same result can be stated by saying that *the angular mean of the second-order l.o.s. projected velocity field weighted by* *the projected density distribution Σ, is equal to 1/3 of the virial velocity V_V^2 of the relevant component* (see eq. [3.28a]). The angular mean of the second order PVT reads

$$2[K_{\text{los}}^{(2)}]_\Omega = -\frac{U+W}{3} + \frac{\ddot{I}}{6},\qquad(8.34)$$

as can be easily checked from eq. (8.32a). Obviously, in this particular case, the same result could be obtained by inserting eq. (8.33) in eq. (7.17b).

9. THE JEANS THEOREM

So far, we have mostly addressed the problem of describing time–dependent stellar systems, with focus on the 3-dimensional space. In this Chapter, we start the discussion of the phase–space properties of stationary, collisionless stellar systems. In particular, we show below that for stationary collisionless stellar systems the distribution function can be expressed in terms of the regular integrals of the motion. The integrals of the motion reveal deep properties of the solution, but their mathematical description can be very complicated and in many cases they may not exist. Integrals of the motion are particularly important when they are available explicitly and refer to potential of astrophysical interest.

9.1 Introduction

In this Chapter we refer to the particularly important case of *stationary systems*. From a physical point of view, the assumption of stationarity is interesting because the majority of stellar systems (elliptical and spiral galaxies, globular clusters) appear to be described by (more or less) regular morphologies, smooth surface brightness distributions, and (relatively) simple stellar kinematics. All these properties suggest that the objects are in a stationary (or quasi–stationary) state. From a mathematical point of view stationarity is also interesting, because a powerful theorem (the *Jeans theorem*) on a general property of the distribution function of stationary systems is available. Many general results on the dynamics of collisionless stationary stellar systems then follow from basic arguments. In Chapter 11 we will present (in very simple cases) the results that can be obtained by combining the generality of the method of Moments with the strength of the Jeans theorem.

In order to arrive at the main result of this Chapter, i.e., the proof of the Jeans theorem, it is necessary to review some basic facts about the *integrals of the motion* (sometimes also called *first integrals*) for a generic ODE. A vast literature is available on the subject. Here we will limit our discussion to what is needed in the present context, but hopefully we will be able to remain on solid grounds and avoid some pitfalls that may be met in more ambitious presentations. We start by presenting the following

Definition 9.1 [Stationary system] *A collisionless stellar system with distribution function f is said to be* stationary *if and only if*

$$
\begin{cases}
\dfrac{\partial f}{\partial t} = 0 \quad \forall t \quad \text{i.e.} \\[2mm]
\left< \mathbf{v}, \dfrac{\partial f}{\partial \mathbf{x}} \right> - \left< \dfrac{\partial \phi_{\mathrm{T}}}{\partial \mathbf{x}}, \dfrac{\partial f}{\partial \mathbf{v}} \right> = 0,
\end{cases}
\tag{9.1}
$$

In a stationary system, all macroscopic quantities derived from f (density, streaming velocity, velocity dispersions, etc.) are independent of time; in particular, $\partial \phi_{\mathrm{T}} / \partial t = 0 \ \forall t$. The concept of *integral of the motion* for a generic (stationary) potential ϕ is introduced in the next Section.

9.2 Integrals of the motion

For simplicity, let us consider *autonomous* ODEs only, that is, ordinary differential equations for which the associated vector field is independent of time (see eq. [2.1a] and the following discussion).

Definition 9.2 [Integral of the motion] *Consider the autonomous ODE*

$$\begin{cases} \dot{\mathbf{x}} = \mathbf{W}(\mathbf{x}) \\ \mathbf{x}(0) = \mathbf{x}^0, \end{cases} \tag{9.2}$$

where $\mathbf{x}^0 \in A \subseteq \Re^n$, $\mathbf{W} \in \mathcal{C}^{(r)}(A)$ $(r \geq 1)$, *and with solution* $\mathbf{\Psi} = \mathbf{\Psi}(\mathbf{x}^0; t)$
(see Chapter 2).
I) A function $I : A \mapsto \Re$, $I = I(\mathbf{x})$ *so that* $\forall \mathbf{x}^0 \in A$

$$\frac{dI_{\mathcal{L}}(\mathbf{x}^0; t)}{dt} = 0, \qquad I_{\mathcal{L}} := I[\mathbf{\Psi}(\mathbf{x}^0; t)], \tag{9.3a}$$

is called integral of the motion *for* $\mathbf{W}$ *over A.*
II) If I is an integral of the motion, and $I \in \mathcal{C}^{(r)}(A)$ $(r \geq 1)$, *then I is called*
a regular integral of the motion *for* $\mathbf{W}$ *over A.*
III) If I_i, $(i = 1, ..., k)$ *are regular integrals of the motion over A and, more-*
over, the vectors $\partial I_i / \partial \mathbf{x}$ *are* linearly independent *over A, then the k integrals*
I_i *are said to be* functionally independent *over A.*
IV) If I is a regular integral of the motion for $\mathbf{W}$ *and satisfies the Implicit*
Function theorem over A, at least with respect to one variable x_i, *i.e., if*
$I^0 = I(\mathbf{x}^0)$, *then a function* h_i *exists so that*

$$x_i = h_i(I^0, x_1, ..., x_{i-1}, x_{i+1}, ..., x_n), \tag{9.3b}$$

and

$$I(x_1, ..., x_{i-1}, h_i, x_{i+1}, ..., x_n) = I^0, \qquad \forall \mathbf{x}^0 \in A; \tag{9.3c}$$

then I is an isolating integral of the motion *for* $\mathbf{W}$ *over A with respect to* x_i.

By definition, isolating integrals are necessarily regular, but the converse is
not true. In the above definition, all the various integrals of the motion are
global, i.e., they exist over all A. When they exist only in local sense, i.e.,
when A is a small region around $\mathbf{x}^0$, they are called *local integrals of the*
motion. Note that from the Implicit Function theorem, regular and isolating
integrals are *locally* equivalent.

   In the theory of ODEs (see, e.g., [1.6]), the following concept is of great
importance

Definition 9.3 [Jacobi–Lie integrability] *An autonomous ODE as that in*
eq. (9.2) *is said to be* integrable in the sense of Jacobi–Lie *if and only if*

$n-1$ *functionally independent, global, isolating integrals exist. Otherwise, it is called* non–integrable.

Let us discuss the meaning of this definition. When the assumptions given in eqs. (9.3abc) are verified, without loss of generality we can write for a generic initial condition $\mathbf{x}^0 \in A$, and globally over A:

$$\begin{cases} x_i = h_i(I_1^0, I_2^0, ..., I_{n-1}^0, x_n), \quad (i = 1, ..., n-1) \\ \dfrac{dx_n}{dt} = W_n(h_1, h_2, ..., h_{n-1}, x_n) = w_n(I_1^0, I_2^0, ..., I_{n-1}^0, x_n). \end{cases} \tag{9.4}$$

Here the value of I_i^0 are fixed by the initial condition $\mathbf{x}^0$. The remaining differential equation for x_n can be integrated (at least in principle) by separation of variables (obviously, the functions h_i in eqs. [9.4] and [9.3b] are not the same!). In other words, the Jacobi–Lie integrability condition is the technical translation of the old–fashioned name of *integration by quadratures*. Note that a generic autonomous ODE can well possess global, unique solutions *without* being integrable in the restricted sense of Jacobi–Lie. In other words, *a non–integrable ODE (in the Jacobi–Lie sense) is characterized by a vector field that does not possess a sufficient number $(n-1)$ of functionally independent global isolating integrals of the motion.*

The request of globality is particularly important. In fact, the following result (that we will not prove) holds independently of the integrability of the vector field $\mathbf{W}$:

Theorem 9.4 [Flow–Box theorem] *Let* $\mathbf{x}^0 \in A \subseteq \Re^n$ *an ordinary point for the vector field* $\mathbf{W} \in C^{(r)}(A)$ $(r \geq 1)$, *i.e.* $\mathbf{W}(\mathbf{x}^0) \neq 0$. *Then, a neighborhood* U_0 *of* $\mathbf{x}^0$ *and a function* $\phi \in C^{(r)}(U_0)$ $(r > 1)$ *exist so that, with the coordinate transformation*

$$\mathbf{y} := \phi(\mathbf{x}), \tag{9.5a}$$

eq. (9.2) transforms into

$$\begin{cases} \dot{y}_i = U_{ik}(\mathbf{x}^0)W_k(\mathbf{x}) = 0 \quad (i = 1, ..., n-1) \\ \dot{y}_n = U_{nk}(\mathbf{x}^0)W_k(\mathbf{x}) = 1 \\ \mathbf{y}^0 = \phi(\mathbf{x}^0), \end{cases} \tag{9.5b}$$

where

$$U_{ik}(\mathbf{x}) := \frac{\partial \phi_i}{\partial x_k}. \tag{9.5c}$$

PROOF   See, e.g., [2.3] and [4.1].   ◁

The solution of eq. (9.5b) is straightforward, i.e.,

$$\begin{cases} y_i = y_i^0 = \phi_i(\mathbf{x}^0) & (i = 1, ..., n-1) \\ y_n = t; \end{cases} \tag{9.6}$$

this explains the origin of the other name for this result, "rectification theorem". The specific form of ϕ is determined by $\mathbf{W}$. Note that, according to the previous definitions, the first $n-1$ components of $\phi(\mathbf{x}^0)$ are regular, *local* integrals of the motion for $\mathbf{W}$; in other words, the regular local integrals of the motion are just the ϕ_i for $i = 1, ..., n-1$. Finally, note that the previous theorem holds for any ODE associated with a well–behaved vector field, independently of the property of Jacobi–Lie integrability. A natural question arises: could we use the initial conditions as global integrals of the motion for any ODE? After all, the n coordinates of the initial conditions are conserved for any time along the orbit, not only locally! Unfortunately, as we will see below, the answer is negative. A clear understanding of this negative result is important, because many inaccurate (and even incorrect) statements can be found in the literature on this point.

We will follow two different lines of reasoning in order to prove the previous statement, the first based on a qualitative–geometrical approach, and the second on more technical grounds. We will then present a simple example of the complex topological nature of initial conditions when considered as global integrals of the motion, even in Jacobi–Lie integrable systems.

The basic idea in the geometric argument is that of the *time elimination* between the n components of the general solution. In particular, following the results obtained in Chapter 2, when the solution exists (and is unique), it follows that $\mathbf{x}^0 = \Psi(\mathbf{x}; -t)$. The possibility of using initial conditions as global integrals of the motion is equivalent to the time elimination *for all times* between $n-1$ components of Ψ. In fact, let us assume that a function h_n exists so that

$$t = h_n(\mathbf{x}, x_n^0); \tag{9.7a}$$

then, after substitution

$$h_i(\mathbf{x}, x_n^0) := \Psi_i[\mathbf{x}, -h_n(\mathbf{x}, x_n^0)] = x_i^0, \tag{9.7b}$$

for $i = 1, ...n-1$, and so the set of h_i would be a family of global integrals of the motion for $\mathbf{W}$. The problem here is that there is no reason to expect that the function h_n be defined better than in a local sense. This remark is true even for systems that are actually Jacobi–Lie integrable! The following simple example clarifies this point.

Example 9.5 [Two–dimensional harmonic oscillator]
Let us consider the standard 2-dimensional harmonic oscillator, described by

$$\begin{cases} \ddot{x} = -\lambda^2 x, \\ \ddot{y} = -\mu^2 y, \end{cases} \tag{9.8a}$$

where λ and μ are two positive constants. As is well known, two isolating, global integrals of the motion are the two energies, i.e.,

$$I_x := \frac{\dot{x}^2}{2} + \frac{\lambda^2 x^2}{2}, \quad I_y := \frac{\dot{y}^2}{2} + \frac{\mu^2 y^2}{2}. \tag{9.8b}$$

The general solution, in terms of initial conditions, is given by

$$\begin{cases} x(t) = x_0 \cos(\lambda t) + \dfrac{\dot{x}_0}{\lambda} \sin(\lambda t) = \dfrac{\sqrt{2I_x^0}}{\lambda} \cos(\lambda t - \varphi_x^0), \\[4mm] y(t) = y_0 \cos(\mu t) + \dfrac{\dot{y}_0}{\mu} \sin(\mu t) = \dfrac{\sqrt{2I_y^0}}{\mu} \cos(\mu t - \varphi_y^0), \end{cases} \tag{9.8c}$$

where, for the x component,

$$\varphi_x^0 := \mathrm{Arccos}\left(\frac{\lambda x_0}{\sqrt{2I_x^0}}\right), \tag{9.8d}$$

and $0 \leq \mathrm{Arccos}(z) \leq \pi$ is the *principal determination* of arccos[9.1]. By inverting the first of eq. (9.8c), one obtains

$$\lambda t = \varphi_x^0 + 2\pi k - \mathrm{Arccos}\left[\frac{\lambda x(t)}{\sqrt{2I_x^0}}\right], \tag{9.8e}$$

where $k = 0, \pm 1, \pm 2, ...$, and $t = 0$ for $k = 0$. From eq. (9.8e) the impossibility of a global inversion with respect to time for $-\infty < t < \infty$ is apparent. But this is only the *first* inconvenience. After substitution of eq. (9.8e) in the second of eq. (9.8c), one obtains

$$y(t) = \frac{\sqrt{2I_y^0}}{\mu} \cos\left\{ \frac{2\pi\mu k}{\lambda} - \frac{\mu}{\lambda}\mathrm{Arccos}\left[\frac{\lambda x(t)}{\sqrt{2I_x^0}}\right] + \frac{\mu}{\lambda}\varphi_x^0 - \varphi_y^0 \right\}. \tag{9.8f}$$

[9.1] Note that with this determination we are in fact assuming that $\dot{x}_0 > 0$ and $\dot{y}_0 > 0$.

Now, for a given $x(t)$, if μ and λ are linearly dependent over the rationals, $y(t)$ assumes a finite number of values (and in general more than one). The situation is worse for μ and λ rationally independent, in which case for a given $x(t)$ the set of the corresponding $y(t)$ is dense.

The general impossibility of using the initial conditions as regular, global integrals of the motion can be shown formally by means of the Flow–Box theorem. According to Theorem 9.4, close to every non–equilibrium point of a generic ODE given by eq. (9.2), the y_i coordinates remain constant for $i = 1, ..., n-1$, while $y_n = t$. The possibility to use initial conditions as global (regular) integrals of the motion is equivalent to requiring the validity of this rectification not only locally, but also globally. This is formally equivalent to require that the transformation U_{ik}, considered as a function of $\mathbf{x}$, satisfies globally the following system

$$\begin{cases} U_{ik}(\mathbf{x})W_k(\mathbf{x}) = 0 & (i = 1, ..., n-1) \\ U_{nk}(\mathbf{x})W_k(\mathbf{x}) = 1. \end{cases} \tag{9.9a}$$

This system of equations determine the functions $U_{ij}(\mathbf{x})$. Now, from eq. (9.5c) the following differential 1–forms are associated with U_{ij}:

$$dy_i := U_{ik}(\mathbf{x})dx_k = 0. \tag{9.9b}$$

In order to define a function (and so, in order to be global, regular integrals of the motion, because on each solution $dy_i = 0$), the functions y_i also need to be *path independent* (i.e., holonomic). But this can be true if and only if the vectors fields U_{ik} are globally exact, i.e.,

$$\frac{\partial U_{il}(\mathbf{x})}{\partial x_m} = \frac{\partial U_{im}(\mathbf{x})}{\partial x_l} \tag{9.9c}$$

for $i = 1, ..., n$ and $l \neq m = 1, ..., n$, a condition that cannot be expected to be satisfied for general vector fields $\mathbf{W}$. This is the reason why initial conditions cannot be used more than locally (i.e., for small t) as regular integrals of the motion. In other words, for non–integrable ODEs (in the sense of Jacobi– Lie), the initial conditions do not define global holonomic coordinates. In contrast, in Jacobi–Lie integrable systems, the global vector field generated by the Flow–Box theorem is exact, and so

$$U_{ik}(\mathbf{x})dx_k = dI_i(\mathbf{x}), \quad (i = 1, ..., n-1). \tag{9.9d}$$

In conclusion, the following passage from a remarkably clear book [1.6] is worth being recalled: "... *every non–integrable flow in phase–space has $n-1$ time–independent global conservation laws,..., but they are singular: during a finite time $0 \leq t < t_1$ the flow can be parallelized via a Lie transformation (eq. [9.9a]), but at time t_1 one of the conservation laws has a singularity and so there is no parallelization by that particular transformation for $t \geq t_1$. According to the Flow–Box theorem, however, there is nothing special about the time t_1: the flow can again be parallelized locally for $t \geq t_1$, and the same argument applies again. This means that the singularities of the conservation laws of a non–integrable flow must be like branch cuts or phase singularities: moveable and arbitrary, like the international date line in attempt to impose linear time on circular time. This leads to the following picture of a non–integrable flow: integrability holds piecewise within infinitely many time intervals $0 \leq t < t_1$, $t_1 \leq t < t_2$, ..., but at the times t_i a singularity of a conservation law is reached and the clock must be reset.*"

How do the previous results apply to vector fields derived from Hamiltonian functions? First of all, for Hamiltonian vector fields it is useful to define a new class of operators on the regular integrals of the motion (if they exist).

Definition 9.6 [Poisson brackets] *For a given Hamiltonian system with $H = H(\mathbf{q}, \mathbf{p})$, $\Gamma = \Re^{2n}$, the Poisson bracket of two regular functions $U, V : \Gamma \mapsto \Re$ is defined as*

$$[U, V] := \sum_{i=1}^{n} \left(\frac{\partial U}{\partial q_i} \frac{\partial V}{\partial p_i} - \frac{\partial U}{\partial p_i} \frac{\partial V}{\partial q_i} \right), \qquad (9.10a)$$

The Poisson brackets satisfy the following relations, the proof of which is left to the reader (or see, e.g., [1.1], [1.3], [1.5], [1.6]):

Theorem 9.7 [Lie algebra of Poisson brackets]
Let U, V, Z be regular functions and a, b constants. Then
I) $[U, V] = -[V, U]$ (antisymmetry)
II) $[aU + bV, Z] = a[U, Z] + b[V, Z]$ (bilinearity)
III) $[UV, Z] = U[V, Z] + V[U, Z]$ (Leibniz rule)
IV) $[U, [V, Z]] + [Z, [U, V]] + [V, [Z, U]] = 0$ (Jacobi identity)

Moreover, for $i, j = 1, ..., n$
V) $[p_i, p_j] = [q_i, q_j] = 0$ $[q_i, p_j] = \delta_{ij}$.

Finally, for any regular function $f = f(\mathbf{q}, \mathbf{p}; t)$, along the phase–flow

VI) $\dfrac{Df}{Dt} = \dfrac{\partial f}{\partial t} + [f, H].$

Points I)-II)-IV) above define a Lie algebra.

Of great importance is the following

Definition 9.8 [Involution] *Two regular functions U, V are said to be in involution if*

$$[U, V] = 0. \tag{9.10b}$$

Note that from this definition and from point VI) above it follows that a time independent function U is a regular integral of the motion for H if and only if $[U, H] = 0$, i.e., if and only if U is in involution with the Hamiltonian. Moreover, using IV) above with $U = H$, one obtains that if V and Z are two regular integrals of the motion for H, so is $[V, Z]$. Finally, for an autonomous system H is an integral of the motion, for $[H, H] = 0$. The importance of the concept of involution is made clear by the following fundamental

Theorem 9.9 [Liouville] *Let $H = H(\mathbf{q}, \mathbf{p})$ a Hamiltonian function, $\Gamma = \Re^{2n}$. Then, the vector field generated by H is Jacobi–Lie integrable if and only if*
I) there exist n global isolating (and so regular) integrals of the motion I_i,
II) I_i are functionally independent over the phase space,
III) they are in involution over the phase–space, i.e., $[I_i, I_j] = 0$ for $i, j = 1, ..., n$.
A Hamiltonian Jacobi–Lie integrable system is also said to be completely canonically integrable.

PROOF   See, e.g., [1.1], [1.2], [1.3], [1.5].   ◁

Thus, the special (symplectic) nature of the ($2n$-dimensional) vector fields originated in phase–space by Hamiltonian functions reduces the number of independent integrals of the motion required for the Jacobi–Lie integrability (when they are in involution!) from $2n - 1$ to n. Finally, note that when the Hamilton–Jacobi equation is *separable*, the constants of separation are global isolating integrals of the motion. It is important to emphasize that separable systems are only a (proper) subset of completely canonically integrable systems, i.e., there exist Hamiltonian systems for which the Liouville theorem is satisfied but cannot be separated in any coordinate systems (for a discussion of this point see, e.g. [1.2] and [1.6]).

9.3 The Jeans theorem

Unfortunately, no general method able to determine whether a given system is completely canonically integrable is presently known. Usually, the "simplest" way to find integrals of the motion makes use of known (more or less evident) symmetries of the system (E. Noether)[9.2].

For example, let us assume ϕ_T to be the potential of a gravitating system. Then

- If the potential is time–independent, $\partial\phi_T/\partial t = 0$, then $H = E$ is a global isolating integral.
- If the potential is time–independent and spherically symmetric, $\phi_T = \phi_T(r)$, then $H = E$ and $\mathbf{L}$ are global isolating integrals.
- If the potential is time–independent and axisymmetric, $\phi_T = \phi_T(R, z)$, then $H = E$ and L_z are global isolating integrals.

How do the above results apply to stationary collisionless stellar systems? We have the following

Theorem 9.11 [Jeans theorem] *Let f be the distribution function of a stationary collisionless stellar system. Then f depends on the phase–space coordinates only through the regular integrals of the motion of ϕ_T, i.e.,*

$$f = f(I_1, I_2, ...). \tag{9.11}$$

PROOF   From eq. (9.1) and the discussion following eq. (9.10b), f is a regular integral of the motion. Conversely, if $f = f(I_1, I_2, \ldots, I_k)$, then

$$\frac{Df}{Dt} = \sum_{i=1}^{k} \frac{\partial f}{\partial I_i} \frac{DI_i}{Dt} = 0,$$

because $DI_i/Dt = 0$ for $i = 1, 2, \ldots, k$, and so the CBE is verified.   ◁

With the aid of this powerful result, we are now able to explore some exact models for stationary collisionless stellar systems.

[9.2] In the N-body problem, in addition to the 7 classical integrals E, $\mathbf{P}$ (or $\mathbf{V}_{CM}$), and $\mathbf{L}$, available in every inertial frame of reference S_0 (see Theorem 3.2), there are the 3 *constants of the motion* $\mathbf{R}_{CM} - t\mathbf{V}_{CM} = \mathbf{R}_{CM}(0)$. These constants become integrals in the subset of inertial frames of reference moving at velocity $\mathbf{V}_{CM}$, where $\mathbf{R}_{CM} = \mathbf{R}_{CM}(0)$. For a definition of constant of the motion see, e.g., [5.3].

10. THE CONSTRUCTION OF A GALAXY MODEL. I
FROM f TO ρ

After characterizing the general properties of stationary collisionless stellar systems, we now proceed, in this Chapter, to the construction of equilibrium models. More specifically, we formulate and discuss here the so–called *direct problem of collisionless stationary Stellar Dynamics*. The present approach is best suited for systems where statistical and/or dynamical arguments can lead to a plausible *ansatz* for the form of the underlying distribution function, expressed in terms of the relevant integrals of the motion. In the absence of such ansatz and in the presence of specific request on the macroscopic profiles (such as density and velocity dispersion profiles) a different approach is often followed, which will be described in the next two chapters.

10.1 Introduction

The direct problem of collisionless stationary Stellar Dynamics consists in the assignment of N distribution functions, dependent on a set of isolating integrals of the motions and some free parameters. For each distribution function the Poisson equation must be solved (in the total potential), and the properties of the system obtained are then studied (for example, by comparison with observations). Particular care must be paid to the concept of *component* of a stellar system: with the term component we mean each part of the system that is characterized by specific properties, i.e., chemical composition, age, mass–to–light ratio, etc. In principle, the "number" of physically distinct components can be "infinite", when the DF depends on some continuous parameter, e.g., in the presence of a continuous mass spectrum. In this case the total DF is obtained by integration over such parameter. For simplicity, in the following we will consider only the discrete case. A basic requirement to be satisfied by the DF associated with each component is that it has to be non–negative over the phase–space. When this request is satisfied, each component is said to be *consistent*. In other words, a *self–consistent* stellar system is a self–gravitating system for which each component is consistent. For reasons that will be made clear in the following, it is useful to work with *relative* energies and potentials. These quantities are defined as follows:

Definition 10.1 [Relative energy and potential] *The* relative potential Ψ_T *and* relative energy *per unit mass* $\mathcal{E}$*, for a system in which* $\ddot{\mathbf{x}} = -\partial\phi_T/\partial\mathbf{x}$ *and* $E = \phi_T + ||\mathbf{v}||^2/2$*, are*

$$\Psi_T := -\phi_T, \qquad \mathcal{E} := -E = \Psi_T - \frac{||\mathbf{v}||^2}{2}. \tag{10.1}$$

The concept of self–consistent multi–component system is formally introduced by:

Definition 10.2 [Collisionless self–consistent multi–component system] *For* $k = 1, ..., n$ *let* $f_k = f_k(\mathbf{x}, \mathbf{v}; t)$ *be regular functions. The function*

$$f := \sum_{k=1}^{n} f_k \tag{10.2a}$$

is said to be the DF of a collisionless self–consistent multi–component system *if and only if for* $i = 1, ..., n$ *and* $\forall (\mathbf{x}, \mathbf{v}; t) \in \Gamma \times \Re$

$$\frac{\partial f_k}{\partial t} + < \mathbf{v}, \frac{\partial f_k}{\partial \mathbf{x}} > + < \frac{\partial \Psi_T}{\partial \mathbf{x}}, \frac{\partial f_k}{\partial \mathbf{v}} >= 0, \tag{10.2b}$$

$$f_k \geq 0, \tag{10.2c}$$

$$\triangle \Psi_T = -4\pi G \rho_T, \tag{10.2d}$$

where

$$\rho_T := \sum_{k=1}^{n} \rho_k = \sum_{k=1}^{n} \int_{\Re^3} f_k d^3 \mathbf{v}. \tag{10.2e}$$

In this case each component f_k *is said to be a* consistent component.

10.2 The direct problem of Stellar Dynamics

The direct problem of collisionless stationary Stellar Dynamics can be formulated as follows:

Definition 10.3 [Direct problem of collisionless stationary Stellar Dynamics] *Let* $\Gamma = \Re^6$. *Let* $I_i : \Gamma \mapsto \Re$, $(i = 1, ..., m)$ *be regular* functions [i.e., at least $\mathcal{C}^{(1)}$] functionally independent *over* Γ, *with*[10.1] $I_1 := \mathcal{E} = \Psi_T - ||\mathbf{v}||^2/2$. *Moreover, let*

$$f_k = f_k(\lambda_k, I_i) : A_k \times \Gamma \mapsto \Re, \quad (k = 1, ..., n), \tag{10.3a}$$

where for $k = 1, ..., n$, *the* vector parameter $\lambda_k \in A_k \subseteq \Re^{d_k}$, *with* d_k *a natural number.*

For $k = 1, ..., n$, *and for each* λ_k, *the following subset of* Γ *is associated with* f_k:

$$\Omega_k := \{(\mathbf{x}, \mathbf{v}) \in \Gamma : f_k \geq 0, \lambda_k \in A_k\}, \tag{10.3b}$$

with f_k regular *over* Ω_k. *For* $k = 1, ..., n$, *for each* λ_k, *and for each* $\mathbf{x} \in \Omega_k$,

$$\Omega_{k\mathbf{x}} := \{\mathbf{v} : (\mathbf{x}, \mathbf{v}) \in \Omega_k, \lambda_k \in A_k\}, \tag{10.3c}$$

is called velocity section *of* Ω_k *at* $\mathbf{x}$.

[10.1] Note that, due to the time independence of Ψ_T, the (isolating) integral $I_1 = \mathcal{E}$ is *always* available.

The density of each component is functionally obtained as

$$\rho_k(\mathbf{x}, \Psi_T) := \int_{\Omega_{k\mathbf{x}}} f_k d^3\mathbf{v}. \tag{10.3d}$$

After defining the following three sets

$$\Lambda_+ := \{\boldsymbol{\lambda}_k \in A_k : \Omega_k \neq \emptyset \text{ for } k = 1, ..., n\}, \tag{10.4a}$$

$$\Lambda_\Psi := \{\boldsymbol{\lambda}_k \in A_k : \triangle\Psi_T = -4\pi G\rho_T \text{ has positive solution over } \Re^3\}, \tag{10.4b}$$

$$\Lambda_I := \{\boldsymbol{\lambda}_k \in A_k : I_i \text{ are global regular integrals of the motion for } \Psi_T\}, \tag{10.4c}$$

the direct problem is solvable if and only if $\Lambda_\exists := \Lambda_+ \cap \Lambda_\Psi \cap \Lambda_I \neq \emptyset.$

Remark 10.3a

If the direct problem is solvable, from the Jeans theorem the functions f_k are (stationary) distribution functions for each $\boldsymbol{\lambda}_k \in \Lambda_\exists$, and the stellar system described by $f = \sum_{k=1}^n f_k$ is a collisionless, stationary self–consistent multi–component system. For practical purposes, the starting point in concrete applications is the assumption of a particular geometry for the system (i.e., for Ψ_T), so that the functions I_i are determined by symmetry considerations. In this framework, $\Lambda_I = \cup_{k=1}^n A_k$, and so the problem reduces to the determination of the sets Λ_+ and Λ_Ψ, and to the subsequent determination of Ψ_T (and ρ_k). Finally, the sets Ω_k are in general proper sub–sets of Γ. The *boundaries* $\partial\Omega_k$ then necessarily depend on $\mathbf{x}$ and $\mathbf{v}$ through the integrals of the motion I_i only.

Before embarking in the detailed description of a particularly simple family of stellar systems constructed following the direct problem, we discuss some general properties of systems that are generated by elementary choices of integrals of the motion and functional forms for f_k.

Theorem 10.4 [Isotropic systems] *Let* $I_1 := \mathcal{E} = \Psi_T - ||\mathbf{v}||^2/2$, *and*

$$f := \begin{cases} f(\mathcal{E}) \geq 0, & \text{if } \mathcal{E} > 0, \\ 0, & \text{if } \mathcal{E} \leq 0. \end{cases} \tag{10.5a}$$

Then

$$\rho = 4\pi \int_0^{\Psi_T} \sqrt{2(\Psi_T - \mathcal{E})} f(\mathcal{E}) d\mathcal{E}. \tag{10.5b}$$

Moreover, for $i \neq j = 1, 2, 3$

$$\overline{v_i} = 0, \quad \overline{v_i v_j} = 0, \tag{10.5c}$$

and

$$\rho \sigma_{ii}^2 = \frac{4\pi}{3} \int_0^{\Psi_T} [2(\Psi_T - \mathcal{E})]^{3/2} f(\mathcal{E}) d\mathcal{E}, \tag{10.5d}$$

for $i = 1, 2, 3$.

PROOF   From eq. (10.3b), $\Omega = \{(\mathbf{x}, \mathbf{v}) \in \Gamma : \Psi_T - ||\mathbf{v}||^2/2 \geq 0\}$, and according to eq. (10.3c) the velocity sections $\Omega_{k\mathbf{x}}$ are spheres of radius $\sqrt{2\Psi_T(\mathbf{x})}$. As direct consequence of the geometry of $\Omega_{k\mathbf{x}}$ the natural coordinates in velocity–space are the spherical ones, i.e.,

$$v_\mathbf{x} := v \sin \lambda \, \cos \mu, \quad v_\mathbf{y} := v \sin \lambda \, \sin \mu, \quad v_\mathbf{z} := v \cos \lambda, \tag{10.5e}$$

with $0 \leq v = ||\mathbf{v}|| \leq \sqrt{2\Psi_T}$, $0 \leq \lambda < \pi$, and $0 \leq \mu < 2\pi$. From eq. (10.3d) one then obtains

$$\rho = \int_{\Omega_{k\mathbf{x}}} f d^3\mathbf{v} = 4\pi \int_0^{\sqrt{2\Psi_T}} f v^2 dv,$$

where the integration variables are changed according to eq. (10.5e), with

$$d^3\mathbf{v} = v^2 \sin \lambda \, dv \, d\lambda \, d\mu;$$

changing variable, from v to $\mathcal{E}$, one obtains

$$dv = -\frac{d\mathcal{E}}{\sqrt{2(\Psi_T - \mathcal{E})}},$$

and so eq. (10.5b) is recovered because $0 \leq \mathcal{E} \leq \Psi_T$. With a similar procedure it is trivial to show that eq. (10.5c) holds. Finally,

$$\rho \sigma_{ii}^2 = \int_0^{\sqrt{2\Psi_T}} f v^2 dv \int_0^\pi \sin \lambda \, d\lambda \int_0^{2\pi} v_i^2 d\mu,$$

and after integration over the angular variables eq. (10.5d) is recovered.   ◁

Under the hypothesis of Theorem 10.4, the quantity ρ depends on $\mathbf{x}$ only through Ψ_T, [i.e., $\rho = \rho(\Psi_T)$]; streaming motions cannot be present and the velocity dispersion tensor is *globally isotropic* (and so $\overline{v_i^2} = \sigma_{ii}^2 = \sigma^2$ for $i = 1, 2, 3$).

An important (and perhaps unexpected) consequence can be derived by imposing self–consistency (through the Poisson equation [10.2c]) on a generic isotropic system,

$$\triangle \Psi = -4\pi G \rho(\Psi_T). \tag{10.5f}$$

In fact, all stationary systems (independently of their specific form) admit the orbital energy as global isolating integral of the motion, and one could feel free to imagine the existence of (isotropic) stationary systems without special symmetries, characterized by distribution functions that depend only on the energy, because of the absence of other global isolating integrals of the motion. Curiously, a remarkable result (see, e.g., [6.12]) states that the *only* positive and non truncated solutions for eq. (10.5f) (with $\Psi = \Psi_T$) are *spherically symmetric*, and so one obtains:

Theorem 10.5 *A non spherically symmetric, non truncated one–component stellar system, with no other regular integral of the motion in addition to $\mathcal{E}$, cannot be stationary (and so even $\mathcal{E}$ is not an integral of motion).*

PROOF   In fact, we assume the system to be stationary, so that $\mathcal{E}$ is an isolating integral of the motion, and non–spherical, with no other regular integrals of the motion. From the Jeans theorem, necessarily $f = f(\mathcal{E})$, but from the result contained in [6.12] the Poisson equation does not admit acceptable solutions. The contradiction implies that one of the assumptions is wrong, i.e., the system cannot be stationary.   ◁

As a consequence, in order to construct one–component non–spherical self–consistent stellar systems, at least *two* regular integrals of the motion are needed. Obviously, all the results derived in Theorem 10.4 apply to spherical stellar systems.

Theorem 10.6 [Anisotropic spherical systems] *Assume the total potential to be spherically symmetric, and $I_1 := \mathcal{E} = \Psi_T - ||\mathbf{v}||^2/2$, $I_2 := ||\mathbf{L}||^2 = L^2$ (where $\mathbf{L} = \mathbf{x} \wedge \mathbf{v}$ is the angular momentum per unit mass). Moreover, let*

$$f := \begin{cases} f(\mathcal{E}, L^2) \geq 0, & \text{if } \mathcal{E} > 0, \\ 0, & \text{if } \mathcal{E} \leq 0. \end{cases} \tag{10.6a}$$

Then

$$\rho = \frac{4\pi}{r^2} \int_0^{\Psi_T} d\mathcal{E} \int_0^{r\sqrt{2(\Psi_T - \mathcal{E})}} \frac{f(\mathcal{E}, L^2) L dL}{\sqrt{2(\Psi_T - \mathcal{E}) - L^2/r^2}}. \tag{10.6b}$$

Moreover,

$$\overline{v_r} = \overline{v_\vartheta} = \overline{v_\varphi} = \overline{v_r v_\vartheta} = \overline{v_r v_\varphi} = \overline{v_\vartheta v_\varphi} = 0, \tag{10.6c}$$

and

$$\rho \sigma_r^2 = \frac{4\pi}{r^2} \int_0^{\Psi_T} d\mathcal{E} \int_0^{r\sqrt{2(\Psi_T - \mathcal{E})}} f(\mathcal{E}, L^2) \sqrt{2(\Psi_T - \mathcal{E}) - \frac{L^2}{r^2}} L dL, \tag{10.6d}$$

$$\rho\sigma_t^2 = \frac{4\pi}{r^4} \int_0^{\Psi_T} d\mathcal{E} \int_0^{r\sqrt{2(\Psi_T - \mathcal{E})}} \frac{f(\mathcal{E}, L^2)L^3 dL}{\sqrt{2(\Psi_T - \mathcal{E}) - L^2/r^2}}. \tag{10.6e}$$

PROOF   From eq. (10.3b) $\Omega = \{(\mathbf{x}, \mathbf{v}) \in \Gamma : \Psi_T - ||\mathbf{v}||^2/2 \geq 0\}$, and according to eq. (10.3c) the velocity sections $\Omega_{k\mathbf{x}}$ are spheres of radius $\sqrt{2\Psi_T(\mathbf{x})}$. In this case the natural coordinates in configuration space are the spherical ones, (r, ϑ, φ) (with $r = ||\mathbf{x}||$), and the associated velocity components are $(v_r, v_\vartheta, v_\varphi)$, with $d^3\mathbf{v} = dv_r dv_\vartheta dv_\varphi$. In fact, $L = rv_t$ ($v_t^2 := v_\vartheta^2 + v_\varphi^2$), and $v^2 = v_r^2 + v_t^2$: thus v_ϑ and v_φ appear in the DF only through v_t. As a direct consequence of the geometry of $\Omega_{k\mathbf{x}}$ the natural coordinates in velocity–space are the spherical ones, i.e.,

$$v_\vartheta := v \sin \lambda \, \cos \mu, \; v_\varphi := v \sin \lambda \, \sin \mu, \; v_r := v \cos \lambda, \tag{10.6f}$$

with $0 \leq v = ||\mathbf{v}|| \leq \sqrt{2\Psi_T}$, $0 \leq \lambda < \pi$, and $0 \leq \mu < 2\pi$. Note that $v_t = v \sin \lambda$. From eq. (10.3d) one then obtains

$$\rho = 4\pi \int_0^{\sqrt{2\Psi_T}} v^2 dv \int_0^{\pi/2} f \sin \lambda \, d\lambda,$$

where the integration variables are changed according to eq. (10.6f), with

$$d^3\mathbf{v} = v^2 \sin \lambda \, dv \, d\lambda \, d\mu;$$

changing variables from (v, λ) to $(\mathcal{E}, L)$, one obtains

$$dv d\lambda = \frac{d\mathcal{E} dL}{r\sqrt{2(\Psi_T - \mathcal{E})}\sqrt{2(\Psi_T - \mathcal{E}) - L^2/r^2}},$$

and so eq. (10.6b) is recovered, because $0 \leq \mathcal{E} \leq \Psi_T$, $0 \leq L \leq r\sqrt{2(\Psi_T - \mathcal{E})}$. With a similar procedure, it is possible to show that eq. (10.6c) holds. Finally,

$$\rho\sigma_r^2 = 4\pi \int_0^{\sqrt{2\Psi_T}} v^4 dv \int_0^{\pi/2} f \sin \lambda \, \cos^2 \lambda \, d\lambda,$$

$$\rho\sigma_t^2 = 4\pi \int_0^{\sqrt{2\Psi_T}} v^4 dv \int_0^{\pi/2} f \sin^3 \lambda \, d\lambda,$$

and after integration over the angular coordinates eqs. (10.6de) are recovered.   ◁

Under the hypothesis of Theorem 10.6, the density ρ depends explicitly both on r and Ψ_T, [i.e., $\rho = \rho(r, \Psi_T)$]; streaming motions cannot be present, and $\sigma_\varphi^2 = \sigma_\vartheta^2 = \sigma_t^2/2$. At variance with the case discussed in Theorem 10.4, the velocity dispersion tensor is *anisotropic*, and the associated velocity dispersion ellipsoids are axisymmetric around the radial direction. In order to characterize this type of anisotropy, the function

$$\beta(r) := 1 - \frac{\sigma_t^2/2}{\sigma_r^2} \tag{10.6g}$$

is introduced. When $\beta(r) = 0$ the system is isotropic (at r), when $\beta(r) > 0$ the system is *radially anisotropic* (i.e., the velocity dispersion ellipsoid at r is prolate), and when $\beta(r) < 0$ the system is *tangentially anisotropic* at r (i.e., the velocity dispersion ellipsoid at r is oblate).

Two special systems are included in Theorem 10.6, the totally radial and totally tangential spherical systems. The radial case is obtained from eqs. (10.6b) and (10.6de), when $f(\mathcal{E}, L^2) = h(\mathcal{E})\delta(L^2)$. With elementary transformations,

$$\rho = \frac{2\pi}{r^2} \int_0^{\Psi_T} \frac{h(\mathcal{E})d\mathcal{E}}{\sqrt{2(\Psi_T - \mathcal{E}) - L^2/r^2}}, \tag{10.6h}$$

$$\rho\sigma_r^2 = \frac{2\pi}{r^2} \int_0^{\Psi_T} h(\mathcal{E})\sqrt{2(\Psi_T - \mathcal{E})}d\mathcal{E}. \tag{10.6i}$$

Obviously, $\sigma_t^2 = 0$. The tangential case is obtained, for any $\Psi_T(r)$, with a DF of the form

$$f(r, v_r, v_t) = \frac{\rho(r)}{2\pi v_c(r)}\delta(v_r)\delta[v_t - v_c(r)], \tag{10.6l}$$

where $v_c^2(r) = -r\, d\Psi_T/dr$ is the *circular velocity* at radius r. The identities $\sigma_r^2 = 0$ and $\sigma_t^2 = v_c^2$ are easily checked[10.2].

Theorem 10.7 [Anisotropic OM spherical systems] *Assume that the total potential is spherically symmetric, and* $I_1 := \mathcal{E} = \Psi_T - ||\mathbf{v}||^2/2$, $I_2 := ||\mathbf{L}||^2 = L^2$. *Moreover, let*

$$Q := \mathcal{E} - \frac{L^2}{2r_a^2}, \tag{10.7a}$$

[10.2] Note that in this case the DF is *not* written *explicitly* in terms of the integrals of the motions $\mathcal{E}$ and L^2, because $r = r(\mathcal{E})$ from

$$\mathcal{E} = \Psi_T(r) - \frac{r}{2}\frac{d\Psi_T(r)}{dr}.$$

and

$$f := \begin{cases} f(Q) \geq 0, & \text{if } Q > 0, \\ 0, & \text{if } Q \leq 0, \end{cases} \tag{10.7b}$$

where r_{a} is a free parameter called anisotropy radius *(for reasons that will be clear in the following). Then*

$$\rho = \frac{4\pi}{1 + r^2/r_{\mathrm{a}}^2} \int_0^{\Psi_T} \sqrt{2(\Psi_T - Q)} f(Q) dQ. \tag{10.7c}$$

Moreover eq. (10.6c) holds, and

$$\rho\sigma_{\mathrm{r}}^2 = \frac{1}{3} \frac{4\pi}{1 + r^2/r_{\mathrm{a}}^2} \int_0^{\Psi_T} [2(\Psi_T - Q)]^{3/2} f(Q) dQ, \tag{10.7d}$$

$$\rho\sigma_{\mathrm{t}}^2 = \frac{2}{3} \frac{4\pi}{(1 + r^2/r_{\mathrm{a}}^2)^2} \int_0^{\Psi_T} [2(\Psi_T - Q)]^{3/2} f(Q) dQ. \tag{10.7e}$$

PROOF   As in Theorem 10.6 the natural coordinates in configuration space are the spherical ones. In fact, from eq. (10.3b) $\Omega = \{(\mathbf{x}, \mathbf{v}) \in \Gamma : \Psi_T - \|\mathbf{v}\|^2/2 - L^2/2r_{\mathrm{a}}^2 = \Psi_T - v_{\mathrm{r}}^2/2 - v_{\mathrm{t}}^2(1 + r^2/r_{\mathrm{a}}^2)/2 \geq 0\}$, and according to eq. (10.3c) the velocity sections $\Omega_{k\mathbf{x}}$ are rotation ellipsoids around v_{r}. As a direct consequence of the geometry of $\Omega_{k\mathbf{x}}$ the natural coordinates in velocity–space are the elliptical ones, i.e.,

$$v_{\vartheta} := \frac{v \sin \lambda \, \cos \mu}{\sqrt{1 + r^2/r_{\mathrm{a}}^2}}, \quad v_{\varphi} := \frac{v \sin \lambda \, \sin \mu}{\sqrt{1 + r^2/r_{\mathrm{a}}^2}}, \quad v_{\mathrm{r}} := v \cos \lambda, \tag{10.7f}$$

with $0 \leq v \leq \sqrt{2\Psi_T}$, $0 \leq \lambda \leq \pi$, and $0 \leq \mu < 2\pi$. Note that $v_{\mathrm{t}} = v \sin \lambda / \sqrt{1 + r^2/r_{\mathrm{a}}^2}$. From eq. (10.3d) one then obtains

$$\rho = \frac{2\pi}{1 + r^2/r_{\mathrm{a}}^2} \int_0^{\sqrt{2\Psi_T}} v^2 dv \int_0^{\pi} f \sin \lambda \, d\lambda = \frac{4\pi}{1 + r^2/r_{\mathrm{a}}^2} \int_0^{\sqrt{2\Psi_T}} f v^2 dv,$$

where the integration variables are changed according to eq. (10.7f), with

$$d^3\mathbf{v} = \frac{v^2 \sin \lambda}{1 + r^2/r_{\mathrm{a}}^2} \, d\lambda \, d\mu;$$

changing variables from v to Q, one obtains

$$dv = -\frac{dQ}{\sqrt{2(\Psi_T - Q)}},$$

and so eq. (10.7c) is recovered because $0 \leq Q \leq \Psi_T$. With a similar procedure it is trivial to show that eq. (10.6c) holds. Finally,

$$\rho\sigma_{\mathrm{r}}^2 = \frac{2\pi}{1 + r^2/r_{\mathrm{a}}^2} \int_0^{\sqrt{2\Psi_T}} v^4 f dv \int_0^{\pi} \sin \lambda \cos^2 \lambda \, d\lambda,$$

and

$$\rho\sigma_t^2 = \frac{2\pi}{(1 + r^2/r_a^2)^2} \int_0^{\sqrt{2\Psi_T}} v^4 f dv \int_0^\pi \sin^3 \lambda \, d\lambda,$$

and after integration over the angular variables eqs. (10.7de) are recovered. ◁

The particular parameterization discussed in the above Theorem was introduced by Ossipkov ([6.26]) and Merritt ([6.23]), and so these models are often called *OM systems*. Stellar systems described by eqs. (10.7ab) are obviously a subset of those presented in Theorem 10.6. They do not present streaming velocities, while their velocity dispersion tensor is anisotropic with two different components, σ_r^2 and σ_t^2. The relation between radial and tangential velocity dispersions in OM systems is very simple (see also Chapter 11). From eq. (10.6g) we find

$$\beta(r) = \frac{r^2}{r^2 + r_a^2}. \tag{10.7g}$$

As a consequence, in OM systems the velocity dispersion tensor is isotropic at the center *independently* of the particular value of r_a. When $r_a \to \infty$, $\beta(r) = 0 \; \forall r \geq 0$, i.e, the system is globally isotropic, according to the fact that now $Q = \mathcal{E}$. For fixed r_a the velocity dispersion tensor becomes more and more radially anisotropic for $r \to \infty$, while inside r_a the velocity dispersion tensor is more and more isotropic. This is the reason of the name of anisotropy radius for the parameter r_a. Finally, there is a striking similarity between eq. (10.7c) and eq. (10.5b) for isotropic systems, except for the radial factor in front of the integral.

A related family of models is obtained from eqs. (10.7ab) if we refer to the parameter

$$Q := \mathcal{E} + \frac{L^2}{r_a^2}. \tag{10.7h}$$

Most of the previous discussion remains formally unchanged, with the difference that in this case the plus sign in radial factors outside the integrals becomes a minus, and so

$$\beta(r) = -\frac{r^2}{r_a^2 - r^2}. \tag{10.7i}$$

The resulting velocity dispersion tensor is still isotropic in the center, but now increasingly tangentially anisotropic with increasing radius. In contrast, this parameterization, at variance with the radially anisotropic case, can be applied only to *truncated systems*, provided that the limitation of $r_a > r_t$, where r_t is the so called *truncation radius*, i.e., $\rho(r) = 0$ for $r \geq r_t$. Therefore, for this class of models, only an upper limit on the tangential anisotropy is allowed. Because of this limitation, this class of models is not commonly used.

Theorem 10.8 [Anisotropic Cuddeford's spherical systems] *Assume that the total potential is spherically symmetric, and* $I_1 := \mathcal{E} = \Psi_T - ||\mathbf{v}||^2/2$, $I_2 := ||\mathbf{L}||^2 = L^2$. *Moreover, let* Q *be defined by eq. (10.7a), and*

$$f := \begin{cases} L^{2\alpha} h(Q) \geq 0, & \text{if } Q > 0, \\ 0, & \text{if } Q \leq 0, \end{cases} \tag{10.8a}$$

with $\alpha > -1/2$. *Then,*

$$\rho = \frac{4\pi r^{2\alpha}}{(1+r^2/r_a^2)^{\alpha+1}} \frac{\sqrt{\pi}}{2} \frac{\Gamma(\alpha+1)}{\Gamma(\alpha+3/2)} \int_0^{\Psi_T} [2(\Psi_T - Q)]^{\alpha+1/2} h(Q) dQ. \tag{10.8b}$$

Moreover eq. (10.6c) holds, and

$$\rho \sigma_r^2 = \frac{4\pi r^{2\alpha}}{(1+r^2/r_a^2)^{\alpha+1}} \frac{\sqrt{\pi}}{4} \frac{\Gamma(\alpha+1)}{\Gamma(\alpha+5/2)} \int_0^{\Psi_T} [2(\Psi_T - Q)]^{\alpha+3/2} h(Q) dQ, \tag{10.8c}$$

$$\rho \sigma_t^2 = \frac{4\pi r^{2\alpha}}{(1+r^2/r_a^2)^{\alpha+2}} \frac{\sqrt{\pi}}{2} \frac{\Gamma(\alpha+2)}{\Gamma(\alpha+5/2)} \int_0^{\Psi_T} [2(\Psi_T - Q)]^{3/2} h(Q) dQ. \tag{10.8d}$$

PROOF   The proof is similar to that of Theorem 10.7. Changing coordinates according to eq. (10.7f), one obtains

$$\rho = \frac{2\pi r^{2\alpha}}{(1+r^2/r_a^2)^{\alpha+1}} \int_0^{\sqrt{2\Psi_T}} v^{2\alpha+2} h dv \int_0^\pi \sin^{2\alpha+1} \lambda \, d\lambda,$$

and so eq. (10.8b) is recovered. The convergence of the angular integral requires $\alpha > -1/2$. Moreover

$$\rho \sigma_r^2 = \frac{2\pi r^{2\alpha}}{(1+r^2/r_a^2)^{\alpha+1}} \int_0^{\sqrt{2\Psi_T}} v^{2\alpha+4} h dv \int_0^\pi \sin^{2\alpha+1} \lambda \, \cos^2 \lambda \, d\lambda,$$

$$\rho \sigma_t^2 = \frac{2\pi r^{2\alpha}}{(1+r^2/r_a^2)^{\alpha+2}} \int_0^{\sqrt{2\Psi_T}} v^{2\alpha+4} h dv \int_0^\pi \sin^{2\alpha+3} \lambda \, d\lambda,$$

and, after integration over the angular variables, eqs. (10.8cd) are recovered.   ◁

This is a generalization of the OM systems (see [6.7]), and the velocity dispersion anisotropy is described by

$$\beta(r) = \frac{r^2 - \alpha r_a^2}{r^2 + r_a^2}. \tag{10.8e}$$

Therefore, in Cuddeford's systems, at the center the velocity dispersion is tangentially anisotropic (for $-1/2 < \alpha < 0$) or radially anisotropic (for $\alpha > 0$) *independently* of the particular value of r_a, while for $r \to \infty$ the anisotropy is totally radial, as in OM systems. When $r_a \to \infty$, $\beta(r) = -\alpha$ $\forall r \geq 0$, i.e, the system is everywhere tangentially anisotropic. For $\alpha = 0$ one recovers the OM models. Finally, for Cuddeford's systems, a related family of models (for truncated systems) can also be obtained using eq. (10.7h). The determination of $\beta(r)$ in this case and its interpretation are left to the reader.

Theorem 10.9 [Two–integral axisymmetric systems] *Assume that the total potential is axisymmetric [i.e., in standard cylindrical coordinates, $\Psi_T = \Psi_T(R,z)$, independent of the azimuthal angle φ], and $I_1 := \mathcal{E} = \Psi_T - ||\mathbf{v}||^2/2$, $I_2 := L_z$, (where L_z is the component of the angular momentum along the z-axis). Moreover, let*

$$f := \begin{cases} f_+(\mathcal{E}, L_z) + f_-(\mathcal{E}, L_z) \geq 0, & \text{if } \mathcal{E} > 0, \\ 0, & \text{if } \mathcal{E} \leq 0, \end{cases} \tag{10.9a}$$

where $f_\pm$ are the even (odd) component of the DF with respect to L_z, i.e.,

$$f_\pm(\mathcal{E}, L_z) = \frac{f(\mathcal{E}, L_z) \pm f(\mathcal{E}, -L_z)}{2}. \tag{10.9b}$$

Then

$$\rho = \frac{4\pi}{R} \int_0^{\Psi_T} d\mathcal{E} \int_0^{R\sqrt{2(\Psi_T - \mathcal{E})}} f_+(\mathcal{E}, L_z) dL_z. \tag{10.9c}$$

Moreover

$$\overline{v_R} = \overline{v_z} = \overline{v_R v_z} = \overline{v_R v_\varphi} = \overline{v_z v_\varphi} = 0, \tag{10.9d}$$

while

$$\rho \overline{v_\varphi} = \frac{4\pi}{R^2} \int_0^{\Psi_T} d\mathcal{E} \int_0^{R\sqrt{2(\Psi_T - \mathcal{E})}} f_-(\mathcal{E}, L_z) L_z dL_z. \tag{10.9e}$$

Finally,

$$\rho \overline{v_\varphi^2} = \frac{4\pi}{R^3} \int_0^{\Psi_T} d\mathcal{E} \int_0^{R\sqrt{2(\Psi_T - \mathcal{E})}} f_+(\mathcal{E}, L_z) L_z^2 dL_z, \tag{10.9f}$$

$$\rho\sigma_m^2 = \frac{4\pi}{R}\int_0^{\Psi_T} d\mathcal{E} \int_0^{R\sqrt{2(\Psi_T - \mathcal{E})}} f_+(\mathcal{E}, L_z)\left[2(\Psi_T - \mathcal{E}) - \frac{L_z^2}{R^2}\right] dL_z. \quad (10.9g)$$

PROOF   From eq. (10.3b) $\Omega = \{(\mathbf{x}, \mathbf{v}) \in \Gamma : \Psi_T - ||\mathbf{v}||^2/2 \geq 0\}$, and according to eq. (10.3c) the velocity sections $\Omega_{k\mathbf{x}}$ are spheres of radius $\sqrt{2\Psi_T(\mathbf{x})}$. In this case the natural coordinates in configuration space are the cylindrical ones (R, φ, z), and the associated velocity components are (v_R, v_z, v_φ), with $d^3\mathbf{v} = dv_R dv_z dv_\varphi$. Note that in the DF v_R and v_z appear only in the combination $v_m^2 = v_R^2 + v_z^2$, because $L_z = Rv_\varphi$ and $v^2 = v_m^2 + v_\varphi^2$; v_m is the so-called meridional velocity. As a direct consequence of the geometry of $\Omega_{k\mathbf{x}}$ the natural coordinates in velocity space are the spherical ones, i.e.,

$$v_R := v\sin\lambda\,\cos\mu, \; v_z := v\sin\lambda\,\sin\mu, \; v_\varphi := v\cos\lambda, \qquad (10.9h)$$

with $0 \leq v = ||\mathbf{v}|| \leq \sqrt{2\Psi_T}$, $0 \leq \lambda < \pi$, and $0 \leq \mu < 2\pi$. Note that $v_m = v\sin\lambda$. From eq. (10.3d) one then obtains

$$\rho = 2\pi \int_0^{\sqrt{2\Psi_T}} v^2 dv \int_0^\pi (f_+ + f_-)\sin\lambda\,d\lambda$$

$$= 4\pi \int_0^{\sqrt{2\Psi_T}} v^2 dv \int_0^{\pi/2} f_+ \sin\lambda\,d\lambda,$$

where the integration variables are changed according to eq. (10.9h), with

$$d^3\mathbf{v} = v^2\sin\lambda\,d\lambda\,d\mu;$$

changing coordinates from (v, λ) to $(\mathcal{E}, L_z)$, one obtains

$$dv d\lambda = \frac{d\mathcal{E}\,dL_z}{R\sqrt{2(\Psi_T - \mathcal{E})}\sqrt{2(\Psi_T - \mathcal{E}) - L_z^2/R^2}},$$

and so eq. (10.9c) is proved, because $0 \leq \mathcal{E} \leq \Psi_T$, $0 \leq L_z \leq R\sqrt{2(\Psi_T - \mathcal{E})}$. With a similar procedure it is trivial to show that eq. (10.9d) holds, and that

$$\rho\overline{v_\varphi} = 2\pi \int_0^{\sqrt{2\Psi_T}} v^3 dv \int_0^\pi (f_+ + f_-)\sin\lambda\,\cos\lambda\,d\lambda$$

$$= 4\pi \int_0^{\sqrt{2\Psi_T}} v^3 dv \int_0^{\pi/2} f_- \sin\lambda\,\cos\lambda\,d\lambda.$$

Finally, after integration over the angular coordinates

$$\rho\overline{v_\varphi^2} = 4\pi \int_0^{\sqrt{2\Psi_T}} v^4 dv \int_0^{\pi/2} f_+ \sin\lambda\,\cos^2\lambda\,d\lambda,$$

$$\rho\sigma_m^2 = 4\pi \int_0^{\sqrt{2\Psi_T}} v^4 dv \int_0^{\pi/2} f_+ \sin^3\lambda \, d\lambda,$$

and eqs. (10.9fg) are proved.   ◁

Only the *even* part of the DF contributes to the density. Furthermore, ρ depends explicitly both on R and Ψ_T, [i.e., $\rho = \rho(R, \Psi_T)$)]. The only admissible streaming motion is along the azimuthal direction, and it is determined uniquely by the *odd* component of the DF. The velocity dispersion ellipsoids are rotationally symmetric around the azimuthal direction, with $\sigma_\varphi^2 = \overline{v_\varphi^2} - \overline{v_\varphi}^2$, and $\sigma_R^2 = \sigma_z^2 = \sigma_m^2/2$. Thus, the systems described here can be either isotropic or azimuthally anisotropic.

A particularly simple application of the above Theorem is obtained by considering

$$f_+ = A|L_z|^{2\alpha}\mathcal{E}^\beta, \tag{10.9i}$$

with $A > 0$, $\alpha > -1/2$ and $\beta > -1$. This choice determines the so–called *Fricke models* (see [6.11]). From eq. (10.9g) one obtains immediately

$$\rho = \frac{A\pi 2^{\alpha+5/2} B(\alpha + 3/2, \beta + 1)}{2\alpha + 1} R^{2\alpha}\Psi_T^{\alpha+\beta+3/2}, \tag{10.9l}$$

where $B(x, y)$ is the complete Beta function, i.e., the density is a combination of powers of radius and potential. Note the striking similarity with the dependence of the DF on angular momentum and energy. The recovery of the analogous expressions for σ_m^2 and σ_φ^2 is left to the reader.

Theorem 10.10 [A special case of two–integral axisymmetric systems] *Assume that the total potential be axisymmetric, and $I_1 := \mathcal{E} = \Psi_T - ||\mathbf{v}||^2/2$, $I_2 := L_z$. Moreover, let*

$$Q := \mathcal{E} - \frac{L_z^2}{2R_a^2}, \tag{10.10a}$$

and

$$f = f_+ := \begin{cases} |L_z|^{2\alpha}h(Q) \geq 0, & \text{if } Q > 0, \\ 0, & \text{if } Q \leq 0, \end{cases} \tag{10.10b}$$

where $\alpha > -1/2$, and R_a is a free parameter called anisotropy radius (*for reasons that will be clear in the following). Then*

$$\rho = \frac{4\pi R^{2\alpha}}{(2\alpha + 1)(1 + R^2/R_a^2)^{\alpha+1/2}} \int_0^{\Psi_T} [2(\Psi_T - Q)]^{\alpha+1/2}h(Q)dQ. \tag{10.10c}$$

Moreover eq. (10.9d) holds, $\overline{v_\varphi} = 0$, and

$$\rho\sigma_\varphi^2 = \frac{4\pi R^{2\alpha}}{(2\alpha+3)(1+R^2/R_a^2)^{\alpha+3/2}} \int_0^{\Psi_T} [2(\Psi_T - Q)]^{\alpha+3/2} h(Q) dQ, \quad (10.10d)$$

$$\rho\sigma_m^2 = \frac{8\pi R^{2\alpha}}{(2\alpha+1)(2\alpha+3)(1+R^2/R_a^2)^{\alpha+1/2}} \int_0^{\Psi_T} [2(\Psi_T - Q)]^{\alpha+3/2} h(Q) dQ. \quad (10.10e)$$

PROOF   As in Theorem 10.9 the natural coordinates in configuration space are the cylindrical ones. In fact, from eq. (10.3b) $\Omega = \{(\mathbf{x}, \mathbf{v}) \in \Gamma : \Psi_T - ||\mathbf{v}||^2/2 - L_z^2/2R_a^2 = \Psi_T - v_m^2/2 - v_\varphi^2(1+R^2/R_a^2)/2 \geq 0\}$, and according to eq. (10.3c) the velocity sections $\Omega_{k\mathbf{x}}$ are rotation ellipsoids around the azimuthal direction. Note that from eq. (10.9e) no streaming motions are associated with eq. (10.10b). As a direct consequence of the geometry of $\Omega_{k\mathbf{x}}$ the natural coordinates in velocity space are the elliptical ones, i.e.

$$v_R := v \sin\lambda \cos\mu, \quad v_z := v \sin\lambda \sin\mu, \quad v_\varphi := \frac{v \cos\lambda}{\sqrt{1+R^2/R_a^2}}, \quad (10.10f)$$

with $0 \leq v \leq \sqrt{2\Psi_T}$, $0 \leq \lambda < \pi$, and $0 \leq \mu < 2\pi$. Note that $v_m = v \sin\lambda$. From eq. (10.3d) one then obtains

$$\rho = \frac{4\pi R^{2\alpha}}{(1+R^2/R_a^2)^{\alpha+1/2}} \int_0^{\sqrt{2\Psi_T}} hv^{2\alpha+2} dv \int_0^{\pi/2} \sin\lambda \cos^{2\alpha}\lambda \, d\lambda$$

$$= \frac{4\pi R^{2\alpha}}{(2\alpha+1)(1+R^2/R_a^2)^{\alpha+1/2}} \int_0^{\sqrt{2\Psi_T}} hv^{2\alpha+2} dv,$$

where the integration variables are changed according to eq. (10.10f), with

$$d^3\mathbf{v} = \frac{v^2 \sin\lambda}{\sqrt{1+R^2/R_a^2}} d\lambda \, d\mu;$$

changing variables from v to Q one obtains

$$dv = -\frac{dQ}{\sqrt{2(\Psi_T - Q)}}$$

and so eq. (10.10c) is recovered because $0 \leq Q \leq \Psi_T$. Note that $\alpha > -1/2$ is required for convergence. Finally

$$\rho\sigma_\varphi^2 = \frac{4\pi R^{2\alpha}}{(1+R^2/R_a^2)^{\alpha+3/2}} \int_0^{\sqrt{2\Psi_T}} hv^{2\alpha+4} dv \int_0^{\pi/2} \sin\lambda \cos^{2\alpha+2}\lambda \, d\lambda,$$

$$\rho\sigma_m^2 = \frac{4\pi R^{2\alpha}}{(1+R^2/R_a^2)^{\alpha+1/2}} \int_0^{\sqrt{2\Psi_T}} hv^{2\alpha+4} dv \int_0^{\pi/2} \sin^3\lambda \cos^{2\alpha}\lambda \, d\lambda.$$

◁

For this family of models

$$\beta(R, z) := 1 - \frac{\sigma_\varphi^2}{\sigma_{\rm m}^2/2} = \frac{R^2 - 2\alpha R_{\rm a}^2}{R^2 + R_{\rm a}^2}, \tag{10.10g}$$

i.e., in these systems, the distribution of anisotropy is independent of z and of the specific form of $h(Q)$. The interpretation of anisotropy for increasing R and for various $\alpha > -1/2$ is left to the reader. A simple case of models described in this example can be obtained by generalizing the Fricke models (eq. [10.9n]), by assuming $h(Q) = Q^\beta$, i.e.,

$$f = A|L_z|^{2\alpha}Q^\beta, \tag{10.10h}$$

where $\beta > -1$. In particular, from eq. (10.10e),

$$\rho = \frac{A\pi 2^{\alpha+5/2}B(\alpha + 3/2, \beta + 1)}{(2\alpha + 1)(1 + R^2/R_{\rm a}^2)^{\alpha+1/2}} R^{2\alpha}\Psi_{\rm T}^{\alpha+\beta+3/2}. \tag{10.10i}$$

10.3 A simple application: stellar polytropes

This is perhaps the simplest case of the direct problem of Stellar Dynamics, namely the construction of the family of so–called *spherically symmetric self–gravitating stellar polytropes*. The starting point is the parametric and isotropic DF

$$f := \begin{cases} A_n \mathcal{E}^{n-3/2}, & \text{if } \mathcal{E} > 0, \\ 0, & \text{if } \mathcal{E} < 0. \end{cases} \tag{10.11a}$$

From eq. (10.5c) one immediately obtains

$$\rho = B_n \Psi^n, \quad B_n = \frac{(2\pi)^{3/2}\Gamma(n - 1/2)A_n}{\Gamma(n + 1)}, \tag{10.11b}$$

from which we may recognize the condition $n > 1/2$ for the existence of solutions. The Poisson equation in spherical coordinates is

$$\frac{1}{r^2}\frac{d}{dr}\left(r^2\frac{d\Psi}{dr}\right) = -4\pi GB_n\Psi^n, \tag{10.11c}$$

with the natural boundary conditions $\Psi(0) = \Psi_0$ and $d\Psi/dr = 0$ at $r = 0$. Two physical scales can be associated with this problem, and eq. (10.11c) can be cast in dimensionless form by introducing

$$s := \frac{r}{r_0}, \quad \varphi := \frac{\Psi}{\Psi_0}, \quad r_0 := \frac{1}{\sqrt{4\pi GB_n\Psi_0}}. \tag{10.11d}$$

The resulting dimensionless equation is the well–known *Lane-Emden* equation, the solutions of which have been investigated extensively (see, e.g., [4.5]):

$$\frac{1}{s^2}\frac{d}{ds}\left(s^2\frac{d\varphi}{ds}\right) = \begin{cases} -\varphi^n, & \text{if } \varphi \geq 0 \\ 0, & \text{if } \varphi \leq 0. \end{cases} \tag{10.11e}$$

Here, only a brief account of the main properties is given. It can be proved that: 1) for $1/2 < n < 5$ the total mass associated with the density distribution is finite, and the density vanishes at a finite truncation radius r_t; 2) for $n = 5$, the density is non truncated, but the mass is still finite; 3) for $n > 5$ the density is non truncated and the mass is infinite; 4) analytical solutions can be found for two values of n, namely $n = 1$ (linear Helmholtz equation) and $n = 5$ (Schuster solution). For $n = 1$ one obtains:

$$\varphi = \begin{cases} \dfrac{\sin(s)}{s}, & \text{if } s < \pi \\ \dfrac{\pi}{s} - 1, & \text{if } s \geq \pi \end{cases} \tag{10.11f}$$

and for $n = 5$

$$\varphi = \frac{1}{\sqrt{1 + s^2/3}}. \tag{10.11g}$$

In both cases the total mass can be found explicitly from the Gauss theorem, namely

$$M = -\lim_{r \to r_t} \frac{r^2}{G}\frac{d\Psi}{dr} = -\frac{\Psi_0 r_0}{G}\lim_{s \to s_t} s^2\frac{d\varphi}{ds}, \tag{10.11h}$$

i.e., $M_5 = \sqrt{3}\Psi_0 r_0/G$ and $M_1 = \pi\Psi_0 r_0/G$. In the astrophysical literature, the $n = 5$ case is also known as the *Plummer model* (see [6.27]). The name of *stellar polytropes* derives from the following formal similarity with *gaseous polytropes*. The equation of hydrostatic equilibrium for a gaseous sphere, with polytropic equation of state $p = K\rho^\gamma$ (where γ is the so–called *polytropic index*), is given by $dp/dr = -\rho d\phi/dr$, and so

$$\rho^{\gamma-1} = \frac{\gamma - 1}{K\gamma}\Psi. \tag{10.11i}$$

Therefore, eq. (10.11i) is formally the same as eq. (10.11b) when $\gamma = 1 + 1/n$, i.e., the density distribution of a stellar polytrope of index n is the same as that of a gaseous polytrope of index $\gamma = 1 + 1/n$.

A special case of polytropic systems is the so–called *isothermal sphere*, obtained from the DF

$$f = \frac{\rho_1}{(2\pi\sigma^2)^{3/2}}\exp\left(\frac{\mathcal{E}}{\sigma^2}\right), \tag{10.12a}$$

for which the velocity dispersion is independent of radius. This case can be interpreted as a polytropic with $n \to \infty$. Other isotropic DFs are obtained by truncating the isothermal sphere when $\mathcal{E} \leq \mathcal{E}_0$ (where $\mathcal{E}_0$ is a *truncation energy*, see also Chapter 12). Important cases are the *King* models (see [6.20]) associated with

$$f = \frac{\rho_1}{(2\pi\sigma^2)^{3/2}} \left[\exp\left(\frac{\mathcal{E}}{\sigma^2}\right) - \exp\left(\frac{\mathcal{E}_0}{\sigma^2}\right) \right], \qquad (10.12b)$$

and the *Wilson* models ([6.32]), with

$$f = \frac{\rho_1}{(2\pi\sigma^2)^{3/2}} \left[\exp\left(\frac{\mathcal{E}}{\sigma^2}\right) - \exp\left(\frac{\mathcal{E}_0}{\sigma^2}\right) - \frac{\mathcal{E} - \mathcal{E}_0}{\sigma^2} \right]. \qquad (10.12c)$$

Anisotropic spherical models can be obtained by modifying the King (or Wilson) models. For example, the *Michie* models (see [6.24]) are obtained for

$$f = \frac{\rho_1}{(2\pi\sigma^2)^{3/2}} \exp\left(-\frac{L^2}{2r_a^2\sigma^2}\right) \left[\exp\left(\frac{\mathcal{E}}{\sigma^2}\right) - \exp\left(\frac{\mathcal{E}_0}{\sigma^2}\right) \right]. \qquad (10.12d)$$

Finally, a special mention is due to the Bertin & Stiavelli f_∞ (spherical) models (see [6.2]), unrelated to the previously mentioned functions, associated with the distribution function

$$f = \begin{cases} A\mathcal{E}^{3/2} \exp(a\mathcal{E} - cL^2), & \text{if } \mathcal{E} \geq 0 \\ 0, & \text{if } \mathcal{E} \leq 0, \end{cases} \qquad (10.12e)$$

which are constructed also on the basis of Statistical Mechanics arguments, and which reproduce very well the density (light) distribution of elliptical galaxies (see also Chapter 11). A complete description of the models listed above can be found in the references listed at the end of the Lecture Notes.

11. THE CONSTRUCTION OF A GALAXY MODEL. II STARTING WITH THE JEANS EQUATIONS

In this Chapter and in the following we discuss a complementary approach, with respect to that presented in Chapter 10, to the construction of stationary, self–consistent collisionless stellar systems. Here we start with a discussion of the Jeans equations, with a procedure that is natural and often adopted in the study of elliptical galaxies under the constraints of observed luminosity and dispersion profiles. The Abel inversion technique is introduced in this context.

11.1 Introduction

One difficulty with the f–to–ρ approach is that there is little control on the resulting profiles of density and other dynamical quantities, even though when available these solutions are of great interest because they represent physically acceptable equilibrium models (not necessarily stable) of stationary, collisionless stellar systems.

Another approach (the so–called ρ–to–f approach) to the construction of models describing stellar systems, based on the Jeans equations (see Chapter 7), is available and can be useful for its ability to relate observationally accessible quantities, such as the streaming velocity and velocity dispersion, to other interesting but directly unaccessible quantities (such as, for example, the potential and the density field of a given galaxy). Unfortunately, the main problem of the ρ–to–f approach is that, in general, the underlying DF cannot be determined uniquely, and, even when it is, under exceptional circumstances, its numerical recovery is a classical example of an "ill–conditioned" problem. Finally, the inversion process does not guarantee that the DF thus obtained is positive definite. In fact, we recall here that all macroscopic properties of a stellar system are *moments* of microscopic functions over the velocity space, and from the mathematical point of view physically acceptable moments can be originated by an unphysical DF. Under special circumstances it is possible to obtain sufficient information about the DF associated with the system, which is indeed remarkable. Thus, in general, even though the approach described in this Chapter is easier to apply than the f–to–ρ approach, the validity of the results obtained is not guaranteed a priori.

One of the most common applications of the ρ–to–f approach can be summarized as follows:

• A specific form for the density ρ is assumed, possibly dependent on a sufficient number of parameters that are to be determined by the projected luminosity profile. In addition, different density components can be allowed, in order to simulate the presence of a dark matter halo, of other stellar components with different ages, chemical composition, mass–to–light ratio, and so on. In general, the mass–to–light ratio Υ of each density component is assumed to be independent of position, but its numerical value can be different for each density component.

• For each density component the associated Jeans equation are solved (usually numerically). Their *closure* is obtained by a specific *ansatz* (which are is to motivate physically), with the hope to reproduce, for example, the observed (projected) velocity fields.

• When possible, the underlying DF is recovered, and its positivity is investigated. In case of negative values, the DF cannot be accepted, and then the model obtained must be discarded. The recovery of the DF is the most difficult step of the ρ–to–f approach, and is often ignored, leaving the physical significance of the entire process under serious concerns.

In this Chapter we focus on the first two points above, and we describe, for simplicity, the main properties of spherical and axisymmetric models. The last point above will be discussed in the next Chapter.

11.2 Spherical density–potential pairs

The first step is the choice of a density distribution reproducing the observed projected density (luminosity) profile. Therefore, the relation between spatial and projected density for spherical systems needs to be clarified. This is accomplished by using a very general result, due to Abel.

Theorem 11.1 [Abel] *Let* $g_1(x) : \Re \mapsto \Re$, $g_2(x) : \Re \mapsto \Re$, $x_m \leq x \leq x_M$. *For* $0 < \alpha < 1$ *let*

$$f_1(x) := \int_{x_m}^{x} \frac{g_1(t)dt}{(x-t)^\alpha}, \tag{11.1a}$$

$$f_2(x) := \int_{x}^{x_M} \frac{g_2(t)dt}{(t-x)^\alpha}. \tag{11.1b}$$

Then,

$$g_1(x) = \frac{\sin(\pi\alpha)}{\pi} \frac{d}{dx} \int_{x_m}^{x} \frac{f_1(t)dt}{(x-t)^{1-\alpha}}, \tag{11.2a}$$

and

$$g_2(x) = -\frac{\sin(\pi\alpha)}{\pi} \frac{d}{dx} \int_{x}^{x_M} \frac{f_2(t)dt}{(t-x)^{1-\alpha}}. \tag{11.2b}$$

PROOF   We only prove eq. (11.2b). The proof of eq. (11.2a) is similar. By inserting eq. (11.1b) in eq. (11.2b) and by exchanging the order of integration, one obtains

$$g_2(x) = -\frac{\sin(\pi\alpha)}{\pi} \frac{d}{dx} \int_{x}^{x_M} g_2(y)dy \int_{x}^{y} \frac{dt}{(t-x)^{1-\alpha}(y-t)^\alpha}.$$

The inner integral equals $\pi/\sin(\pi\alpha)$. After differentiation of the resulting expression, the result is proved.   ◁

Let us now consider a spherically symmetric stellar system, described by a mass density $\rho = \rho(r)$ and a (constant) mass–to–light ratio Υ, where r is the radial spherical coordinate. The *projected mass density* is indicated with $\Sigma = \Sigma(R)$, where R is the radius on the projection plane (see Chapter 8). The mass density and the projected mass density are related to the *light density* and to the *projected light density* respectively as $\rho = \Upsilon \nu$ and $\Sigma = \Upsilon I$; because of these straightforward relations, in the following discussion we use the pair (ρ, Σ). The following result holds:

Theorem 11.2 [Projection and deprojection of spherical systems] *Let*

$$\begin{cases} \rho(r) > 0 & \text{if } 0 \leq r < r_t, \\ \rho(r_t) \geq 0 \\ \rho(r) = 0 & \text{if } r > r_t. \end{cases} \tag{11.3a}$$

$$\begin{cases} \Sigma(R) > 0 & \text{if } 0 \leq R < R_t, \\ \Sigma(R_t) \geq 0 \\ \Sigma(R) = 0 & \text{if } R > R_t. \end{cases} \tag{11.3b}$$

and $r_t = R_t > 0$ be the so–called truncation radius. *When $r_t = R_t = \infty$, the system is said to be* non–truncated. *It follows that:*

$$\Sigma(R) = 2 \int_R^{R_t} \frac{\rho(r) r \, dr}{\sqrt{r^2 - R^2}}, \tag{11.4a}$$

$$\rho(r) = -\frac{1}{\pi} \int_r^{r_t} \frac{d\Sigma(R)}{dR} \frac{dR}{\sqrt{R^2 - r^2}} + \frac{\Sigma(R_t)}{\pi \sqrt{R_t^2 - r^2}}. \tag{11.4b}$$

PROOF   The proof of eq. (11.4a) is trivial. In fact, in view of the spherical symmetry, let the z axis be directed along the l.o.s., and so $r^2 = R^2 + z^2$, where $R^2 := x^2 + y^2$ and $z^2 \leq r_t^2 - R^2$. By definition of projection, after changing the integration coordinate from z to r, one obtains $\Sigma(R) = 2 \int_0^{\sqrt{R_t^2 - R^2}} \rho(R^2 + z^2) dz = 2 \int_R^{R_t} \rho(r) r \, dr / \sqrt{r^2 - R^2}$.

The proof of eq. (11.4b) is more elaborate. In fact, some preliminary work is needed in order to apply eq. (11.1b) to the inversion of eq. (11.4a). This is done by a re-definition of various quantities entering eq. (11.4a), namely, $x := r^2$, $X := R^2$, $Y := R_t^2$, and $U(X) := \Sigma(\sqrt{X})$, $u(x) := \rho(\sqrt{x})$, and so $U(X) = \int_X^Y u(x) dx / \sqrt{x - X}$. The inversion formula (11.2b) can now be applied with $\alpha = 1/2$, obtaining $-\pi u(x) = d \left[\int_x^Y U(X) dX / \sqrt{X - x} \right] / dx = r^{-1} d \left[\int_r^{R_t} \Sigma(R) R \, dR / \sqrt{R^2 - r^2} \right] / dr$, where the last identity is obtained by restoring the original integration variable. An integration by parts of the last integral and a successive differentiation with respect to r complete the proof. ◁

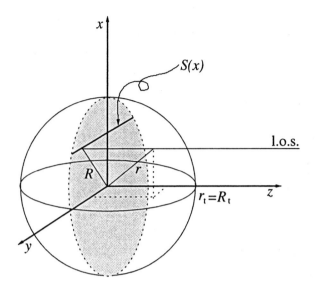

Figure 11.1

Remark 11.2a

Note that if $\rho(r_t) > 0$ is positive and finite, then necessarily $\Sigma(R_t) = 0$. In contrast, when $\Sigma(R_t) > 0$, the associated spatial density is (weakly) divergent at the boundary of the system, $\rho \sim (r_t^2 - r^2)^{-1/2}$.

Before describing the main properties of spherical stellar systems in more detail, we need to establish some basic notation.

Definition 11.3 *For a spherically symmetric stellar system described by the density distribution* $\rho = \Upsilon \nu(r)$ *[and associated* $\Sigma(R) = \Upsilon I(R)$, *where I is the surface brightness profile], the* projected luminosity *inside R is*

$$L_p(R) := 2\pi \int_0^R I(R) R dR. \tag{11.5a}$$

When $I(0)$ is finite, the core radius R_c *is given by*

$$I(R_c) := \frac{I(0)}{2}. \tag{11.5b}$$

When $L_p(R_t)$ is finite, the effective radius R_e *is given by*

$$L_p(R_e) := \frac{L_p(R_t)}{2}. \tag{11.5c}$$

The light contained inside the sphere of radius r is given by

$$L(r) := 4\pi \int_0^r \nu(r)r^2 dr. \tag{11.5d}$$

When $L(r_t)$ is finite, the half–light *radius r_h is given by*

$$L(r_h) := \frac{L(r_t)}{2}. \tag{11.5e}$$

Remark 11.3a

By replacing eq. (11.4a) in eq. (11.5a) and inverting the order of integration, it is easy to show that $L_p(R_t) = L(r_t)$ and thus to prove, in this explicit case, the general result obtained in Chapter 8 (eq. [8.5] with $F = 1$). Moreover, note that, from very simple geometrical arguments, the inequality $R_e \leq r_h$ holds independently of the specific form of $\rho(r)$.

The expressions for the potential ϕ and gravitational self–energy U of a spherical density distribution find important applications because this symmetry is often assumed as a starting point in the construction of galaxy models.

Theorem 11.4 [Potential, gravitational and interaction energies] *For the density distribution given in eq. (11.3a), let $M(r)$ be the mass contained inside the sphere of radius r. Then its gravitational potential is given by*

$$\phi(r) = \begin{cases} -\dfrac{GM(r)}{r} - 4\pi G \displaystyle\int_r^{r_t} \rho(r)r dr, & \text{if } 0 \leq r \leq r_t \\ -\dfrac{GM(r_t)}{r}, & \text{if } r \geq r_t. \end{cases} \tag{11.6a}$$

The associated gravitational self–energy can be written as

$$U = 2\pi \int_0^{r_t} \rho(r)\phi(r)r^2 dr = -\frac{GM(r_t)^2}{2r_t} - \frac{G}{2}\int_0^{r_t} \frac{M(r)^2}{r^2}dr$$

$$= -4\pi G \int_0^{r_t} M(r)\rho(r)r dr. \tag{11.6b}$$

If, in addition, a spherically symmetric external potential $\phi_{ext} = \phi_{ext}(r)$ is present, then the external gravitational energy can be written as

$$U_{ext} = 4\pi \int_0^{r_t} \rho(r)\phi_{ext}(r)r^2 dr$$

$$= -\frac{GM(r_t)M_{ext}(r_t)}{r_t} - G\int_0^{r_t} \frac{M(r)M_{ext}(r)}{r^2}dr. \tag{11.6c}$$

Finally, the interaction energy is given by

$$W = -4\pi G \int_0^{r_t} M_{\text{ext}}(r)\rho(r)r dr. \tag{11.6d}$$

PROOF   Using the Gauss theorem, or integrating the first of eq. (7.3b) over the whole solid angle in spherical coordinates, one obtains $\phi(r) = -G\int_r^\infty M(r)dr/r^2$ (having assumed $\phi(\infty) = 0$). The second of eqs. (11.6a) is then an immediate consequence of this formula, because $M(r) = M(r_t)$ for $r \geq r_t$. For $r \leq r_t$, $\phi(r) = -G\int_r^{r_t} M(t)dt/t^2 - GM(r_t)/r_t$, where the integral can be simplified by changing order of integration: $\int_r^{r_t} M(t)dt/t^2 = 4\pi \int_r^{r_t} dt/t^2 \int_0^t \rho(x)x^2 dx = 4\pi \int_0^r \rho(x)x^2 dx \int_r^{r_t} dt/t^2 + 4\pi \int_r^{r_t} \rho(x)x^2 dx \int_x^{r_t} dt/t^2 = 4\pi \int_r^{r_t} \rho(x)x dx + M(r)/r - M(r_t)/r_t$. This completes the proof. The first identity in eq. (11.6b) is simply the first identity in eq. (7.13a) in spherical coordinates. The second identity can be obtained from the first by integrating by parts with $\rho(r)r^2$ as differential, or from the second identity in footnote 7.4 expressed in spherical coordinates. The third identity can be obtained from eqs. (7.14a) and (7.15c) in spherical coordinates. The first identity in eq. (11.6c) is simply the first identity in eq. (7.13b) in spherical coordinates. The second identity can be obtained from the first by integrating by parts with $\rho(r)r^2$ as differential. Finally, eq. (11.6d) is obtained from eqs. (7.14b) and (7.15d) in spherical coordinates.   ◁

Using the above identities, as an interesting exercise the reader can prove eq. (7.18e) by direct integration.

Related to the projected mass distribution for spherical systems is the following interesting result due to Schwarzschild (see [6.30] and Fig. 11.1):

Theorem 11.5 [Strip–brightness] *Let $I(R) = I(\sqrt{x^2 + y^2})$ be the surface brightness profile associated with the light distribution $\nu(r)$, where $\rho = \Upsilon\nu$. The* strip–brightness $S(x)$ *is defined as the brightness of the strip of $I(R)$ at distance x from $R = 0$. The following relations hold:*

$$S(x) = 2 \int_x^{R_t} \frac{I(R)R dR}{\sqrt{R^2 - x^2}} = 2\pi \int_x^{r_t} \nu(r)r dr, \tag{11.7a}$$

$$\nu(r) = -\frac{1}{2\pi r}\frac{dS(r)}{dr}, \tag{11.7b}$$

$$M(r) = -2\Upsilon \int_0^r \frac{dS(x)}{dx} x dx = -2\Upsilon \left[rS(r) - \int_0^r S(x)dx \right], \tag{11.7c}$$

$$U = -2G\Upsilon^2 \int_0^{R_t} S^2(x)dx. \tag{11.7d}$$

PROOF   We start by proving the first of eq. (11.7a). By definition,
$S(x) := 2 \int_0^{\sqrt{R_t^2-x^2}} I(\sqrt{x^2+y^2})dy = 2 \int_x^{R_t} I(R)RdR/\sqrt{R^2-x^2}$, where the last identity is obtained by changing integration variables from y to R at fixed x. The second
identity in eq. (11.7a) is of more complex derivation. Using eq. (11.4a) and the above
identity, we find

$$S(x) = \int_x^{R_t} \frac{4RdR}{\sqrt{R^2-x^2}} \int_r^{R_t} \frac{\nu(r)rdr}{\sqrt{r^2-R^2}} = \int_r^{R_t} \nu(r)dr \int_x^r \frac{4RdR}{\sqrt{(R^2-x^2)(r^2-R^2)}},$$

where the last identity is obtained by inverting the order of integration; the inner integral
gives 2π, and so the second identity in eq. (11.7a) is proved. Equation (11.7b) is obtained
by differentiation of the second of eq. (11.7a), the first of eq. (11.7c) is obtained by
using eq. (11.7b) and the definition of $M(r)$, and the last identity in eq. (11.7c) by
integrating by parts. Using the last formula of eq. (11.6b) and (11.7b), one obtains $U =$
$-4\pi G \int_0^{r_t} M(r)\rho(r)rdr = 2G\Upsilon \int_0^{r_t} M(r)[dS(r)/dr]dr = -8\pi G\Upsilon^2 \int_0^{r_t} \nu(r)S(r)r^2dr$,
where the last identity is obtained by integrating by parts and using the fact that $S(R_t) =$
0. Using again eq. (11.7b), $U = 4G\Upsilon^2 \int_0^{r_t} S(r)[dS(r)/dr]rdr =$
$= 2G\Upsilon^2 \int_0^{r_t} [dS^2(r)/dr]rdr = -2G\Upsilon^2 \int_0^{r_t} S^2(r)dr$, and so eq. (11.7d) is derived.   ◁

We conclude this short presentation of basic relations between spatial and
projected properties for spherical systems by proving the following

Theorem 11.6 *In a spherical system, for* $r \leq r_t$,

$$\phi(r) = -\frac{2G\Upsilon}{r} \int_0^r S(x)dx, \tag{11.8a}$$

$$\phi(0) = -4G\Upsilon \int_0^{R_t} I(R)dR. \tag{11.8b}$$

PROOF   From the first of eqs. (11.6a) and from eq. (11.7b), we find $\phi(r) =$
$-GM(r)/r + 2\pi G\Upsilon \int_r^{r_t} [dS(x)/dx]dx = -GM(r)/r - 2G\Upsilon S(r) = -(2G\Upsilon/r) \int_0^r S(x)dx$,
where the last identity is obtained using the second of eq. (11.7c). Equation (11.8b) can
be derived in different ways. Perhaps the simplest is that of imposing that $S(0)$ is finite.
In this case, $\lim_{r\to 0} -(2G\Upsilon/r) \int_0^r S(x)dx = -2G\Upsilon S(0) = -4G\Upsilon \int_0^{R_t} I(R)dR$, where
the last identity is obtained using eq. (11.7a).   ◁

Before starting the discussion of the main dynamical properties of spherically
symmetric stellar systems, we list here some of the most used density and
surface brightness profiles. They are relatively simple and yet they represent
reasonably well the density profiles of (early–type) galaxies. Of course, this
list cannot be considered complete.

- de Vaucouleurs "$R^{1/4}$–law" ([6.8])

$$I(R) = I(0) \exp[-7.67(R/R_{\mathrm{e}})^{1/4}], \tag{11.9a}$$

- Sersic "$R^{1/m}$–law" ([6.31])

$$I(R) = I(0) \exp[-b(m)(R/R_{\mathrm{e}})^{1/m}], \tag{11.9b}$$

where $b(m) \sim 2m - 1/3 + 4/405m...$ for $m \to \infty$ (see Example 11.10b).

- Hubble–Reynolds profile ([6.15]-[6.28])

$$I(R) = \frac{I(0)}{(1 + R/R_{\mathrm{H}})^2}, \tag{11.9c}$$

- Modified Hubble profile

$$I(R) = \frac{I(0)}{1 + R^2/R_{\mathrm{H}}^2}, \tag{11.9d}$$

where R_{H} is the core radius.

- King profile ([6.19])

$$I(R) = K \left(\frac{1}{\sqrt{R^2 + R_{\mathrm{K}}^2}} - \frac{1}{\sqrt{R_{\mathrm{t}}^2 + R_{\mathrm{K}}^2}} \right), \quad (R \le R_{\mathrm{t}}) \tag{11.9e}$$

- Power–law

$$\rho(r) = \rho_{\mathrm{c}} \left(\frac{r}{r_{\mathrm{c}}} \right)^{-\gamma}, \quad (0 < \gamma < 3) \tag{11.9f}$$

- β–models

$$\rho(r) = \frac{\rho(0)}{(1 + r^2/r_{\mathrm{c}}^2)^{\beta/2}}, \quad (\beta > 2) \tag{11.9g}$$

- King ([6.21])

$$\rho(r) = \frac{\rho(0)}{(1 + r^2/r_{\mathrm{c}}^2)^{3/2}}, \tag{11.9h}$$

- Plummer ([6.27])

$$\rho(r) = \frac{\rho(0)}{(1 + r^2/r_{\mathrm{c}}^2)^{5/2}}, \tag{11.9i}$$

- γ–models

$$\rho(r) = \frac{(3 - \gamma)M}{4\pi} \frac{r_{\mathrm{c}}}{r^\gamma (r + r_{\mathrm{c}})^{4-\gamma}}, \quad (0 \le \gamma < 3) \tag{11.9j}$$

- Jaffe ([6.18])

$$\rho(r) = \frac{M}{4\pi} \frac{r_{\mathrm{c}}}{r^2 (r + r_{\mathrm{c}})^2}, \tag{11.9k}$$

- Hernquist ([6.14])

$$\rho(r) = \frac{M}{2\pi} \frac{r_{\mathrm{c}}}{r(r + r_{\mathrm{c}})^3}, \tag{11.9l}$$

- Hénon's isochrone ([6.13])

$$\phi(r) = -\frac{GM}{r_{\mathrm{c}} + \sqrt{r^2 + r_{\mathrm{c}}^2}}. \qquad (11.9m)$$

Note that in some of the profiles, the density ρ becomes approximately flat inside some characteristic radius. This scale length is often (and in many cases improperly) called core–radius; usually this radius is *not* the core radius as defined in eq. (11.5b). As anticipated in the Introduction to this Chapter, after choosing a density distribution, the next step in the construction of a model for a stellar system is the solution of the associated Jeans equations, for example with respect to the velocity dispersion. In this case their closure (see Chapter 7) can be obtained by assuming that the (unknown) associated DF depends on certain isolating integrals of the motion (admissible by the model total potential). For example, in spherical models, if we assume $f = f(\mathcal{E}, L^2)$, the Jeans equation to be solved for the density component ρ is

$$\frac{d\rho(r)\sigma_{\mathrm{r}}^2(r)}{dr} + \frac{2\beta(r)}{r}\rho(r)\sigma_{\mathrm{r}}^2(r) = -\rho(r)\frac{d\phi_{\mathrm{T}}(r)}{dr}, \qquad (11.10a)$$

where the definition of $\beta(r)$ is given in eq. (10.6l). For a given choice of the functions $\rho(r)$, $\phi_{\mathrm{T}}(r)$ and $\beta(r)$ in eq. (11.10a) the result of the integration is the function $\sigma_{\mathrm{r}}^2(r)$. Then, from eq. (10.6l), σ_{t}^2 is recovered. The natural boundary condition for eq. (11.10a) is

$$\rho(r_{\mathrm{t}})\sigma_{\mathrm{r}}^2(r_{\mathrm{t}}) = 0, \qquad (11.10b)$$

because, by definition of truncation radius, no stars can cross r_{t} (r_{t} can also be ∞). Because of the linearity of the equation, singular points for the solution may appear only at singular points of the coefficients, i.e., singular points of β, and $\rho d\phi_{\mathrm{T}}/dr$. In physically acceptable models these points may occur only at the center and/or at r_{t}. It is interesting that, in the general case, the solution of eq. (11.10ab) can be cast in explicit form. In fact, eq. (11.10a) is a first order inhomogeneous linear ODE, and its solution is given by the following

Theorem 11.7 *The solution of eqs. (11.ab) is given by*

$$\rho(r)\sigma_{\mathrm{r}}^2(r) = \int_r^{r_{\mathrm{t}}} \rho(x)\frac{d\phi_{\mathrm{T}}(x)}{dx}\exp\left[-2\int_x^r \frac{\beta(y)dy}{y}\right]dx, \qquad (11.11a)$$

$$\sigma_t^2(r) = 2\sigma_r^2(r)[1 - \beta(r)]. \tag{11.11b}$$

PROOF   In order to solve eq. (11.10a), the simplest way is to define the new function $Y(r) := \rho(r)\sigma_r^2(r)$, and to solve the resulting equation with the boundary condition $Y(r_t) = 0$. From the theory of ODEs, the general solution of eq. (11.10a) is given by $Y = Y_0 + Y_1$, where Y_0 is the general solution of the associated homogeneous ODE, and Y_1 is a particular solution of the inhomogeneous ODE. Let us determine Y_0. The integration is trivial, and one obtains

$$Y_0(r) = Y_0(r^0) \exp\left[-2\int_{r^0}^r \frac{\beta(y)dy}{y}\right], \tag{11.11c}$$

where r^0 is a particular radius, for the moment unspecified. In fact, some care is required in the choice of initial condition. The two natural points, i.e., $r^0 = 0$ and $r^0 = r_t$ cannot be chosen at this stage. The choice $r^0 = 0$ cannot be made, because not all admissible functional forms for β result in a convergent integral in eq. (11.11c) (e.g., $\beta = const.$). On the other hand, for $r^0 = r_t$ one has $Y(r^0) = 0 = Y(r)$, which is not the general solution of the homogeneous equation. The function Y_1 is now determined with the method of variation of constants, i.e., we set

$$Y_1(r) := Y_0(r)Y_2(r). \tag{11.11d}$$

After substitution of Y_1 in eq. (11.10a), we are left with the following ODE

$$\frac{dY_2(r)}{dr} = -\frac{\rho(r)}{Y_0(r)}\frac{d\phi_T(r)}{dr},$$

which is easily integrated:

$$Y_2(r) = Y_2(r^0) - \frac{1}{Y_0(r^0)}\int_{r^0}^r \rho(x)\frac{d\phi_T(x)}{dx}\exp\left[2\int_{r^0}^x \frac{\beta(y)dy}{y}\right]dx, \tag{11.11e}$$

where $Y_2(r^0)$ is an arbitrary constant. After substitution of eq. (11.11e) in eq. (11.11d), a rearrangement of $Y_0(r) + Y_1(r)$ shows that, for the choice $Y_2(r^0) = -1$, one obtains $Y(r) = -\int_{r^0}^r \rho(x)[d\phi_T(x)/dx]\exp\left[-2\int_{r^0}^r \beta(y)dy/y + 2\int_{r^0}^x \beta(y)dy/y\right]dx$. Combining the two integrals inside the exponential, it is now possible to choose $r^0 = r_t$, thus proving eq. (11.11a). Note that the boundary condition (11.10b) is satisfied.   ◁

Remark 11.7a

Two particularly simple expressions can be found from eq. (11.11a), namely in the case of global isotropy ($\beta = 0$)

$$\rho(r)\sigma_r^2(r) = G\int_r^{r_t} \frac{\rho(x)M_T(x)}{x^2}dx, \tag{11.12a}$$

and for the case of completely radial orbits ($\beta = 1$)

$$\rho(r)\sigma_r^2(r) = \frac{G}{r^2} \int_r^{r_t} \rho(x)M_T(x)dx. \tag{11.12b}$$

Example 11.7b [Anisotropic OM spherical systems]
In this case, from eqs. (10.7g) and (11.11a), one obtains

$$\rho(r)\sigma_r^2(r) = \frac{G}{r^2 + r_a^2} \int_r^{r_t} \rho(x)M_T(x) \left(1 + \frac{r_a^2}{x^2}\right) dx. \tag{11.12c}$$

When $r_a \to \infty$ and $r_a = 0$, the radial component of the velocity dispersion tensor in isotropic and globally radial stellar systems is obtained.

Example 11.7c [Anisotropic Cuddeford spherical systems]
In this case, from eqs. (10.8e) and (11.11a), one obtains

$$\rho(r)\sigma_r^2(r) = \frac{Gr^{2\alpha}}{(r^2 + r_a^2)^{\alpha+1}} \int_r^{r_t} \rho(x)M_T(x) \left(1 + \frac{r_a^2}{x^2}\right)^{\alpha+1} dx. \tag{11.12d}$$

When $\alpha = 0$, this reduces to the OM solution.

We now derive the projection formula for the second–order velocity dispersion profile associated with spherical models with $f = f(\mathcal{E}, L^2)$. Because in these systems the streaming velocity field vanishes identically, we have

$$\overline{\sigma_1^N}_{\text{los}} = \overline{\sigma_P^N}_{\text{los}} = \overline{v_P^N}_{\text{los}}, \tag{11.13}$$

as discussed in Section 8.2. The following result holds:

Theorem 11.8 *Consider a spherical system described by $f = f(\mathcal{E}, L^2)$, $\rho = \rho(r)$, $\beta = \beta(r)$. Then*

$$\Sigma(R)\overline{v_{\text{Plos}}^2}(R) = 2 \int_R^{R_t} \left[1 - \beta(r)\frac{R^2}{r^2}\right] \frac{\rho(r)\sigma_r^2(r)rdr}{\sqrt{r^2 - R^2}}. \tag{11.14}$$

PROOF   From eq. (8.6), $\Sigma \overline{v_{\text{Plos}}^2} = \int \overline{\rho < \mathbf{n}, \mathbf{v} >^2} d\xi_3$. Given the spherical symmetry, we can assume without loss of generality $\xi_1 = x$, $\xi_2 = y$, and $\xi_1 = z$. As a consequence, $r = \|\xi\|$, $R^2 = \xi_1^2 + \xi_2^2$, and $\mathbf{n} = (0,0,1)$, and so $\Sigma \overline{v_{\text{Plos}}^2} = \int \rho \overline{v_z^2} dz = 2 \int_R^{R_t} \rho(r)\overline{v_z^2}(r)(r^2 - R^2)^{-1/2} dr$, where the last expression is obtained with the change of variable used in eq. (11.4a). Expressing now v_z in terms of the velocity components in spherical coordinates, one obtains $\overline{v_z^2}(r) = \overline{(v_r \cos \vartheta - v_\vartheta \sin \vartheta)^2} = \overline{v_r^2} \cos^2 \vartheta + \overline{v_\vartheta^2} \sin^2 \vartheta$, because $\overline{v_r v_\vartheta} = 0$. Using now eq. (10.6g) $\overline{v_z^2} = \overline{v_r^2}(\cos^2 \vartheta + \sin^2 \vartheta \overline{v_\vartheta^2}/\overline{v_r^2}) = \overline{v_r^2}(1 - \beta \sin^2 \vartheta)$, and from the identity $\sin^2 \vartheta = R^2/r^2$ the result is proved.   ◁

With the aid of Theorem 11.8 for spherical systems we can now check the general result given in eq. (8.33) directly:

Theorem 11.9 [Projected Virial theorem for $f = f(\mathcal{E}, L^2)$ systems]
For a spherical system described by $f = f(\mathcal{E}, L^2)$,

$$[K_{\mathrm{los}}^{(2)}]_\Omega = K_{\mathrm{los}}^{(2)} = \pi \int_0^{R_t} \Sigma(R)\overline{v_{\mathrm{p\,los}}^2}(R)R dR$$

$$= \frac{2\pi}{3} \int_0^{r_t} \rho(r)[\sigma_r^2(r) + \sigma_t^2(r)]r^2 dr = \frac{K}{3}. \qquad (11.15)$$

PROOF   The first identity is a direct consequence of spherical symmetry, and the second identity is the definition of $K_{\mathrm{los}}^{(2)}$ according to eq. (8.22a). The third identity is proved for a generic β using eq. (11.14):

$$\pi \int_0^{R_t} \Sigma(R)\overline{v_{\mathrm{p\,los}}^2}(R)R dR = 2\pi \int_0^{R_t} R dR \int_R^{R_t} \frac{\left[1 - \beta(r)R^2/r^2\right]\rho(r)\sigma_r^2(r)r dr}{\sqrt{r^2 - R^2}}$$

$$= 2\pi \int_0^{r_t} \rho(r)\sigma_r^2(r)r dr \int_0^r \frac{\left[1 - \beta(r)R^2/r^2\right] R dR}{\sqrt{r^2 - R^2}},$$

where the last identity is obtained by inverting the order of integration. Finally, evaluating the inner integral and using eq. (10.6g), one gets $\pi \int_0^{R_t} \Sigma(R)\overline{v_{\mathrm{p\,los}}^2}(R)R dR = (2\pi/3) \int_0^{r_t} \rho\sigma_r^2[3 - 2\beta(r)]r^2 dr = (2\pi/3) \int_0^{r_t} \rho[\sigma_r^2 + \sigma_t^2]r^2 dr = \Pi/6 = K/3$, where the last identity follows from eq. (7.12b).   ◁

As final examples of spherical system, for which many dynamical quantities can be obtained explicitly, let us discuss the case of the OM one–component power–law model (eq. [11.9f]) and the case of the isotropic Sersic model (eq. [11.9b]).

Example 11.10a [Power–law model]
For simplicity, define $s := r/r_c$ in eq. (11.9f), which we rewrite as

$$\rho(r) = \rho_c s^{-\gamma}. \qquad (11.16a)$$

The evaluation of various integrals is trivial but tedious, and so it is left as an exercise for the reader. From eq. (11.5d), the mass inside r is

$$M(r) = M_c \frac{s^{3-\gamma}}{3 - \gamma}, \qquad M_c := 4\pi\rho_c r_c^3, \qquad (11.16b)$$

and so for convergence $\gamma < 3$. The potential is obtained from eq. (11.6a) with $r_t = \infty$:

$$\phi(r) = -\frac{GM_c}{r_c} \frac{s^{2-\gamma}}{(3 - \gamma)(\gamma - 2)}, \qquad \gamma > 2, \qquad (11.16c)$$

where the condition on γ allows us to choose $\phi \to 0$ for $r \to \infty$; note that for $2 < \gamma < 3$ and $r \to 0$ the potential diverges. The projected density is obtained from eq. (11.4a),

$$\Sigma(R) = \rho_c r_c B\left(\frac{1}{2}, \frac{\gamma - 1}{2}\right) \eta^{1-\gamma}, \tag{11.16d}$$

where $\eta := R/r_c$ and $B(x, y)$ is the complete Beta function. The radial component of the velocity dispersion in the OM parameterization is then obtained from eq. (11.12c):

$$\sigma_r^2(r) = \frac{GM_c}{r_c} \frac{s^{2-\gamma}}{2(3-\gamma)(s^2 + s_a^2)} \left(\frac{s^2}{\gamma - 2} + \frac{s_a^2}{\gamma - 1}\right), \tag{11.16e}$$

where $s_a := r_a/r_c$. According to eq. (10.7g), the isotropic and completely radially anisotropic cases are recovered for $r_a \to \infty$ and $r_a = 0$, respectively:

$$[\sigma_r^2(r)]_{\text{iso}} = \frac{GM_c}{r_c} \frac{s^{2-\gamma}}{2(3-\gamma)(\gamma - 1)}, \tag{11.16f}$$

$$[\sigma_r^2(r)]_{\text{rad}} = \frac{GM_c}{r_c} \frac{s^{2-\gamma}}{2(3-\gamma)(\gamma - 2)}. \tag{11.16g}$$

Note how, for this model, the velocity dispersions depend on radius through the potential, i.e., $[\sigma_r^2(r)]_{\text{iso}} = -\phi(r)(\gamma - 2)/2(\gamma - 1)$, $[\sigma_r^2(r)]_{\text{rad}} = -\phi(r)/2$, and

$$\frac{[\sigma_r^2(r)]_{\text{rad}}}{[\sigma_r^2(r)]_{\text{iso}}} = \frac{\gamma - 1}{\gamma - 2} > 1. \tag{11.16h}$$

Therefore, the radial component of the velocity dispersion in the radially anisotropic case is larger than that in the case of orbital isotropy, as expected. Moreover, for fixed r_a and $r \to 0$, we have $\sigma_r^2 \sim [\sigma_r^2(r)]_{\text{iso}}$, while for $r \to \infty$ we find $\sigma_r^2 \sim [\sigma_r^2(r)]_{\text{rad}}$, again as expected from eq. (10.7g). The projected velocity dispersion can be obtained for a generic r_a from eq. (11.14), but, for simplicity, in the following we study the isotropic and totally radial cases only. In the isotropic case ($\beta = 0$ in eq. [11.14]), from eqs. (11.16d) and (11.16f),

$$[\overline{v_{\text{Plos}}^2}(R)]_{\text{iso}} = \frac{GM_c}{r_c} \frac{\eta^{2-\gamma}}{2(3-\gamma)(\gamma - 1)} \frac{B(1/2, \gamma - 3/2)}{B(1/2, \gamma/2 - 1/2)}, \tag{11.16i}$$

and in the completely radial anisotropic case ($\beta = 1$ in eq. [11.14]),

$$[\overline{v_{\text{Plos}}^2}(R)]_{\text{rad}} = \frac{GM_c}{r_c} \frac{\eta^{2-\gamma}}{2(3-\gamma)(\gamma - 2)} \frac{B(3/2, \gamma - 3/2)}{B(1/2, \gamma/2 - 1/2)}. \tag{11.16j}$$

After simplification, it can be shown that

$$\frac{[\overline{v_{\mathrm{plos}}^2}(R)]_{\mathrm{rad}}}{[\overline{v_{\mathrm{plos}}^2}(R)]_{\mathrm{iso}}} = \frac{1}{2(\gamma - 2)}. \qquad (11.16k)$$

Example 11.10b [Sersic models]

The Sersic models are defined as a one–parameter family of stationary, spherical stellar systems, with surface brightness profile generalizing the de Vaucouleurs law. If we define $\eta := R/R_{\mathrm{e}}$, they are given by eq. (11.9b):

$$I(R) = I(0) \exp(-b\eta^{1/m}), \qquad (11.16l)$$

where $I(0)$ is the central surface brightness and R_{e} is the effective radius. Here m is a free parameter, a positive real number, and $b = b(m)$ is a dimensionless function the value of which is determined by the definition of R_{e}. In fact, from eq. (11.5a),

$$L_{\mathrm{p}}(R) = I(0) R_{\mathrm{e}}^2 \frac{2\pi m}{b^{2m}} \gamma(2m, b\eta^{1/m}), \qquad (11.16m)$$

where $\gamma(\alpha, x) = \int_0^x \exp(-t) t^{\alpha-1} dt$ $(\alpha > 0)$ is the *incomplete Gamma function*; by definition, $\gamma(\alpha, \infty) = \Gamma(\alpha)$, where Γ is the *complete Gamma function*, and, for α integer, $\Gamma(\alpha) = (\alpha - 1)!$. The total luminosity associated with eq. (11.16l) is then

$$L = I(0) R_{\mathrm{e}}^2 \frac{2\pi m}{b^{2m}} \Gamma(2m), \qquad (11.16n)$$

and the central potential is obtained from eq. (11.8b):

$$\phi(0) = -G\Upsilon I(0) R_{\mathrm{e}} \frac{4\Gamma(m + 1)}{b^m}. \qquad (11.16o)$$

From eqs. (11.16mn) it follows that $b(m)$ is the solution of the transcendental equation:

$$\gamma(2m, b) = \frac{\Gamma(2m)}{2}. \qquad (11.16p)$$

For example, for $m = 4$ (the de Vaucouleurs law), the numerical solution of eq. (11.16p) is $b(4) \simeq 7.66924944$; it can be proved that for $m \to \infty$ the asymptotic expansion of $b(m)$ is $\sim 2m - 1/3 + 4/405m + 46/25515m^2...$ (see [6.6]).

The luminosity density ν, which is related to the mass density via $\rho(r) = \Upsilon\nu(r)$, is given by eq. (11.4b). With simple substitutions one finds

$$\nu(r) = \frac{I(0)}{R_e} \frac{b^m \alpha^{1-m}}{\pi} C_m(\alpha), \qquad (11.16q)$$

where

$$C_m(\alpha) := \int_1^\infty \frac{\exp(-\alpha t)dt}{\sqrt{t^{2m}-1}}, \qquad (m > 0), \qquad (11.16r)$$

and $\alpha := bs^{1/m}$ is the reduced radial coordinate ($s = r/R_e$). Note that, although the integrand in the function $C_m(\alpha)$ is singular at $t = 1$, this singularity is integrable for any value of m; $C_m(\alpha)$ diverges only for $\alpha = 0$ when $m \leq 1$. However, when $m \leq 1$ the factor α^{1-m} in eq. (11.16q) converges to zero for $\alpha \to 0$, and so a more detailed analysis is required. As a useful exercise, we now determine the leading term of the asymptotic expansions for $\alpha \to 0$ and $\alpha \to \infty$ of the functions $C_m(\alpha)$. For $\alpha \to \infty$ we have

$$C_m(\alpha) \sim \sqrt{\frac{\pi}{2m}} \frac{\exp(-\alpha)}{\sqrt{\alpha}}. \qquad (11.16s)$$

This result can be proved using the standard asymptotic expansion method for real integrals with exponential kernels (see, e.g., [4.3]), with the substitution $x = \sqrt{t^{2m}-1}$ in eq. (11.16r).

For $\alpha \to 0$, three different results are obtained for $m > 1$, $m = 1$, and $m < 1$. In the case $m > 1$, the integral in eq. (11.16r) converges for $\alpha = 0$, and its value can be expressed explicitly:

$$C_m(\alpha) \sim C_m(0) = \frac{B[1/2, (m-1)/2m]}{2m}. \qquad (11.16t)$$

When $m = 1$, eq. (11.16r) reduces to

$$C_m(\alpha) = K_0(\alpha), \qquad (11.16u)$$

where K_0 is the zeroth-order modified Bessel function of the third kind. Finally when $m < 1$, the substitution $x = \alpha t$ in eq. (11.16r) shows that

$$C_m(\alpha) \sim \Gamma(1-m)\alpha^{m-1}. \qquad (11.16v)$$

Thus from eqs. (11.16s) and (11.16stuv) we obtain:

$$\nu(r) \sim \frac{I(0)}{R_e} \sqrt{\frac{b}{2\pi m}} \exp(-bs^{1/m})s^{(1-2m)/2m}, \qquad r \to \infty. \qquad (11.16w)$$

For $m > 1$:

$$\nu(r) \sim \frac{I(0)}{R_e} \frac{B[1/2, (m-1)/2m]}{2mb^{m-1}} \exp(-bs^{1/m}) s^{(1-m)/m}, \quad r \to 0; \quad (11.16x)$$

for $m = 1$

$$\nu(r) \sim \frac{b(1)}{\pi} \ln\left(\frac{2}{bs^{1/2m}}\right), \quad r \to 0; \quad (11.16y)$$

for $0 \leq m < 1$

$$\nu(r) \sim \frac{I(0)}{R_e} \frac{b^m \Gamma(1-m)}{\pi}, \quad r \to 0. \quad (11.16z)$$

It should be noted that for $m > 1$ the density diverges at the origin as $r^{(1-m)/m}$, and therefore the divergence of ρ is worse for higher–m models. For the de Vaucouleurs profile ($m = 4$), the relevant exponent at the origin is $-3/4$.

11.3 Axisymmetric density–potential pairs

Here we present the very basic results about the construction of axisymmetric stellar systems, starting from their density distribution and the Jeans equations. The natural coordinates for these systems are the cylindrical ones, i.e., (R, z, φ). Obviously, axisymmetry means that all model properties do not depend on the azimuthal angle φ.

A very simple procedure that can be adopted in order to generate a special family of axisymmetric models starting from spherical density distributions (for example given by eqs. [11.9f-l]), is based on the substitution

$$r^2 \mapsto m^2 := R^2 + \frac{z^2}{q^2}, \quad (11.17a)$$

in the expression for the density. Here the dimensionless parameter q is responsible for the model flattening. More specifically, for $0 < q < 1$ the density is stratified on *oblate ellipsoids*, for $q = 1$ on spheres, and for $q > 1$ on *prolate ellipsoids*. A surface density distribution that is constant on ellipsoids labeled by

$$m^2 := \frac{x^2}{a^2} + \frac{y^2}{b^2} + \frac{z^2}{c^2}, \quad (11.17b)$$

is called a *homeoid*. The general quadrature formula for the potential associated to homeoidal density distributions $\rho = \rho(m)$ is a classic problem of Mathematical Physics (see, e.g., [4.4], [4.9], [5.3]) and is given by

$$\phi(\mathbf{x}) = -G\pi abc \int_0^\infty \frac{\psi[m(\mathbf{x}, \tau)] d\tau}{\sqrt{(a^2 + \tau)(b^2 + \tau)(c^2 + \tau)}}, \quad (11.17c)$$

where

$$m^2(\mathbf{x}, \tau) := \frac{x^2}{a^2 + \tau} + \frac{y^2}{b^2 + \tau} + \frac{z^2}{c^2 + \tau}, \tag{11.17d}$$

and

$$\psi(m) := 2 \int_m^\infty \rho(m) m \, dm. \tag{11.17e}$$

Of course, other axisymmetric systems, not described by homeoidal distributions, are known. In particular, we can list the

- Miyamoto–Nagai potential–density pair ([6.25])

$$\phi(R, z) = -\frac{GM}{\sqrt{R^2 + \zeta^2}}, \quad \zeta := a + \sqrt{z^2 + b^2}, \tag{11.18a}$$

$$\rho(R, z) = \frac{Mb^2}{4\pi} \frac{aR^2 + (\zeta + 2\sqrt{z^2 + b^2})\zeta^2}{[R^2 + \zeta^2]^{5/2}(z^2 + b^2)^{3/2}}. \tag{11.18b}$$

- Satoh potential–density pair ([6.29])

$$\phi(R, z) = -\frac{GM}{\zeta}, \quad \zeta^2 := R^2 + z^2 + a(a + 2\sqrt{z^2 + b^2}), \tag{11.18c}$$

$$\rho(R, z) = \frac{Mab^2}{4\pi\zeta^3(z^2 + b^2)} \left[\frac{1}{\sqrt{z^2 + b^2}} + \frac{3}{a}\left(1 - \frac{R^2 + z^2}{\zeta^2}\right) \right]. \tag{11.18d}$$

- Binney logarithmic potential–density pair ([5.3])

$$\phi(R, z) = -\frac{v_\infty^2}{2} \ln\left(1 + \frac{R^2 + z^2/q^2}{R_c^2}\right), \tag{11.18e}$$

$$\rho(R, z) = \frac{v_\infty^2}{4\pi G} \frac{(1 + 2q^2)R_c^2 + R^2 + (2 - 1/q^2)z^2}{(R_c^2 + R^2 + z^2/q^2)^2}. \tag{11.18f}$$

As for the spherical case, after the density distribution is assigned, the next step in the construction of a model for an axisymmetric stellar system is the integration of the associated Jeans equations. If we assume that the underlying DF is of the form $f = f(\mathcal{E}, L_z)$, the stationary Jeans equations are given by

$$\frac{\partial \rho \sigma_R^2}{\partial z} = -\rho \frac{\partial \phi_T}{\partial z}, \tag{11.19a}$$

$$\frac{\partial \rho \sigma_R^2}{\partial R} - \rho \frac{\overline{v_\varphi^2} - \sigma_R^2}{R} = -\rho \frac{\partial \phi_T}{\partial R}. \tag{11.19b}$$

The dependence of the DF on $\mathcal{E}$ and L_z implies that $\sigma_R = \sigma_z$. A careful discussion of the boundary condition for the velocity dispersion is needed. *If the system has infinite extent*, then the natural boundary condition is $\rho\sigma_R^2 \to 0$ for $\|\mathbf{x}\| \to \infty$. Truncated systems require that everywhere on the system boundary the *normal* component of the velocity dispersion σ_n vanishes. Let $S(R, z) = 0$ be the surface boundary of the system, where, as usual, $x = R\cos\varphi$, $y = R\sin\varphi$ and $R = \sqrt{x^2 + y^2}$. We now show that the above condition is verified *if and only if* $\sigma_R(= \sigma_z) = 0$ on S. In fact, at any point on the boundary the unit vector is given by

$$\mathbf{n}^{\mathrm{T}} = (n_x, n_y, n_z) = \frac{(S_x, S_y, S_z)}{\sqrt{S_x^2 + S_y^2 + S_z^2}} = \frac{(S_R\cos\varphi, S_R\sin\varphi, S_z)}{\sqrt{S_R^2 + S_z^2}}, \qquad (11.20a)$$

where $S_R := \partial S/\partial R$, $S_x := \partial S/\partial x$ etc., and the normal component of the velocity dispersion is obtained from eq. (7.1a) as $\sigma_n^2 = \overline{<\mathbf{n}, \mathbf{v} - \overline{\mathbf{v}}>^2}$. From simple geometrical arguments, we have $v_x = v_R\cos\varphi - v_\varphi\sin\varphi$ and $v_y = v_R\sin\varphi + v_\varphi\cos\varphi$, and so

$$\sigma_n^2 = n_x^2(\cos^2\varphi\sigma_R^2 + \sin^2\varphi\sigma_\varphi^2) + n_y^2(\sin^2\varphi\sigma_R^2 + \cos^2\varphi\sigma_\varphi^2) + n_z^2\sigma_z^2 +$$

$$2n_xn_y\cos\varphi\sin\varphi(\sigma_R^2 - \sigma_\varphi^2). \qquad (11.20b)$$

From eq. (11.20a) with simple algebraic manipulations, we finally obtain

$$\sigma_n^2 = \frac{S_R^2\sigma_R^2 + S_z^2\sigma_z^2}{S_R^2 + S_z^2} = \sigma_R^2, \qquad (11.20c)$$

and this proves the statement.

The general solution of eqs. (11.19ab) is then easily obtained, by integrating the first of eqs. (11.19a)

$$\rho\sigma_R^2 = \int_z^{z_t(R)} \rho\frac{\partial\phi_{\mathrm{T}}}{\partial z}dz, \qquad (11.21a)$$

where $S[R, z_t(R)] = 0$ (and for infinite systems $z_t = \infty$), and then by performing the differentiation with respect to R on the l.h.s. of eq. (11.19b), thus obtaining $\overline{v_\varphi^2}$. In some cases, the following assumption can be made in order to break the degeneracy between ordered and random motions along the azimuthal direction shown by eq. (11.19b) (where only the sum $\overline{v_\varphi^2} = \overline{v_\varphi}^2 + \sigma_\varphi^2$ appears)

$$\overline{v_\varphi}^2 := k^2(\overline{v_\varphi^2} - \sigma_R^2), \quad 0 \leq k \leq 1, \qquad (11.21b)$$

(see [6.29]), where k is a free parameter. From the above identity one then obtains

$$\sigma_\varphi^2 = k^2\sigma_R^2 + (1 - k^2)\overline{v_\varphi^2} = \sigma_R^2 + (1 - k^2)(\overline{v_\varphi^2} - \sigma_R^2). \qquad (11.21c)$$

Note that for $k = 0$ the azimuthal motions are random, i.e., $\sigma_\varphi^2 = \overline{v_\varphi^2}$, while for $k = 1$ the velocity dispersion tensor is isotropic, i.e., $\sigma_\varphi^2 = \sigma_R^2 (= \sigma_z^2)$, and the term $\overline{v_\varphi^2} - \sigma_\varphi^2$ accounts for ordered motions.

The next step in the model construction is the projection along a l.o.s. (usually tilted with respect to the z axis) of the density and velocity fields, and their comparison with the observational data. This can be accomplished using the results and the methods presented in Chapter 8.

12. THE CONSTRUCTION OF A GALAXY MODEL. III
FROM ρ TO f

In this final Chapter we discuss the last step of the "ρ–to–f" method to the construction of stationary, self–consistent (multi–component) collisionless stellar systems, i.e., the problem of recovering the distribution function under the assumption that the macroscopic profiles (density and velocity dispersion) are known in the entire radial range. It is well known that such reconstruction process is not a priori guaranteed to lead to positive definite distribution functions, and thus in general may fail at its roots. Two fully analytical cases, illustrating explicitly a possible approach to this problem in the case of multi–component stellar systems with a special type of anisotropy profile, are worked out and discussed in detail.

12.1 Introduction

As discussed in Chapter 11, in the ρ–to–f approach one usually one starts from the density profile and then solves the Jeans equations under more or less motivated closure relations. Unfortunately, the fact that the Jeans equations admit a solution that is not obviously unacceptable (for example, a somewhat negative velocity dispersion) is not sufficient to guarantee that the model is viable. In fact, the minimum requirement to be met is that the associated DF of each physically distinct component must be positive definite. A model satisfying this minimal requirement is called a *consistent* model. But the DF is generally difficult to recover, which in many cases makes it impossible to carry out a simple consistency analysis. The general inverse problem of Stellar Dynamics can be formulated as the problem of inverting the *integral equation* (10.3d), for an assigned density distribution. Obviously, this problem can be solved *only* if special assumptions on the functional form of f are made. Inversion techniques are available for integral equations such as eqs. (10.5b), (10.6b), (10.7c), etc. In this Chapter we describe the basic results on the inversion in the simplest cases.

12.2 Recovering the distribution function in special cases

We start by presenting the first result obtained on the inverse problem, namely the inversion formula for spherically symmetric, globally isotropic stellar systems:

Theorem 12.1 [Eddington, 1916] *In a multi—component spherical system, let*

$$f_k(\mathcal{E}) := h_k(\mathcal{E})\Theta(\mathcal{E} - \mathcal{E}_0) \tag{12.1a}$$

be the DF of a globally isotropic density component, where Θ *is the* Heaviside step function, *and* $\mathcal{E}_0 \geq 0$ *the truncation energy. When* $\mathcal{E}_0 = 0$ *the system is non truncated.* Then

$$
\begin{aligned}
h_k(\mathcal{E}) &= \frac{1}{\sqrt{8}\pi^2} \frac{d}{d\mathcal{E}} \int_{\mathcal{E}_0}^{\mathcal{E}} \frac{d\rho_k}{d\Psi_T} \frac{d\Psi_T}{\sqrt{\mathcal{E} - \Psi_T}} \\
&= \frac{1}{\sqrt{8}\pi^2} \left[\int_{\mathcal{E}_0}^{\mathcal{E}} \frac{d^2\rho_k}{d\Psi_T^2} \frac{d\Psi_T}{\sqrt{\mathcal{E} - \Psi_T}} + \frac{1}{\sqrt{\mathcal{E} - \mathcal{E}_0}} \left(\frac{d\rho_k}{d\Psi_T} \right)_{\Psi_T = \mathcal{E}_0} \right]. \tag{12.1b}
\end{aligned}
$$

PROOF   From eq. (10.5b) $\rho_k = 4\pi \int_{\mathcal{E}_0}^{\Psi_T} \sqrt{2(\Psi_T - \mathcal{E})} h_k(\mathcal{E}) d\mathcal{E}$. After differentiation with respect to Ψ_T, the identity above is cast in a form suitable for the integral inversion given in eq. (11.1a), and so the first identity in eq. (12.1b) is proved. The second identity is obtained by integrating by parts the first integral in eq. (12.1b) (with respect to Ψ_T), and then by differentiation (with respect to $\mathcal{E}$).   ◁

Remark 12.1a

Note that the fact that the (proposed) density distribution is positive definite does not guarantee that the recovered DF be also positive definite. The same statement applies even if the velocity dispersions obtained by solving the associated Jeans equations turn out also to be positive definite. If the resulting DF obtained from eq. (12.1b) is negative (somewhere), then the proposed model cannot be physically realized *under the assumptions made on the functional dependence of the DF on the integrals of motion.*

Remark 12.1b

The interested reader can prove the following statements related to Theorem 12.1: 1) for all infinite systems of finite total mass, $\lim_{\Psi_T \to 0} d\rho_k/d\Psi_T = 0$, and so one term in the second identity in eq. (12.1b) is absent; 2) if $h_k \sim (\mathcal{E} - \mathcal{E}_0)^\alpha$ for $\mathcal{E} \to \mathcal{E}_0$, then $\alpha > -1$ to ensure the convergence of the density ρ_k at the edge of the system. When $\alpha > -1$ then $\lim_{\Psi_T \to \mathcal{E}_0} \rho_k = 0$. Moreover $d\rho_k/d\Psi_T \sim (\Psi_T - \mathcal{E}_0)^{\alpha+1/2}$, and so $\lim_{r \to r_t} = d\rho_k/dr = -\infty$ (for $-1 < \alpha < -1/2$), $d\rho_k/dr_t = -cost.$ (for $\alpha = -1/2$), and finally $d\rho_k/dr_t = 0$ (for $\alpha > -1/2$), i.e., the mass density goes smoothly to zero at the edge of the system.

A similar result holds in the case of (multi–component) spherical systems with OM orbital anisotropy (see Theorem 10.7). In this case, for a given density component, we prove the following:

Theorem 12.2 [OM inversion formula] *In a multi–component spherical system, let*

$$f_k(Q_k) := h_k(Q_k)\Theta(Q_k - Q_0) \qquad (12.2a)$$

be the DF of an OM density component. When $Q_0 = 0$ the system is non truncated. Then

$$h_k(Q_k) = \frac{1}{\sqrt{8}\pi^2} \frac{d}{dQ_k} \int_{Q_0}^{Q_k} \frac{d\varrho_k}{d\Psi_T} \frac{d\Psi_T}{\sqrt{Q_k - \Psi_T}} =$$

$$= \frac{1}{\sqrt{8}\pi^2} \left[\int_{Q_0}^{Q} \frac{d^2\varrho_k}{d\Psi_T^2} \frac{d\Psi_T}{\sqrt{Q_k - \Psi_T}} + \frac{1}{\sqrt{Q_k - Q_0}} \left(\frac{d\varrho_k}{d\Psi_T} \right)_{\Psi_T = Q_0} \right], \quad (12.2b)$$

where

$$\varrho_k(r) := \left(1 + \frac{r^2}{r_{ak}^2} \right) \rho_k(r), \quad (12.2c)$$

and $Q_k := \mathcal{E} - L^2/2r_{ak}^2$.

PROOF   The proof is similar to that of the Eddington inversion. The only difference is that the density ϱ is modified by the radial factor appearing in the r.h.s. of eq. (10.7c). ◁

A similar formula can be obtained for the tangentially anisotropic OM case (see Theorem 10.7 and following comments). Other inversion formulae, based on modifications of the basic Abel inversion formula, can be obtained for the Cuddeford systems described in Theorem 10.8. Again, the recovered functions are not guaranteed to be positive definite. A detailed discussion of why the DF can be negative is postponed to the next two Sections, where some analytical cases are worked out explicitly.

The problem of recovering the DF for axisymmetric systems, i.e., the inversion of eqs. (10.9ce), is much more difficult than that of the spherically symmetric case. In general, it is assumed that the elimination of z is possible between $\rho(R, z)$ and $\Psi_T(R, z)$, and so we can derive an expression of the form $\rho = \rho(R, \Psi_T)$. The known methods of inversion are summarized (without technical discussions, that can be found in the relevant references) in the following list:

• **Fricke method**
If $\rho(R, \Psi_T)$ is expressed as a *finite* or *infinite* series of terms $R^a \Psi_T^b$, then eqs. (10.9ij) can be used. The DF is then given by a sum or a series of functions of the type of eq. (10.9i). The proof of the convergence of the obtained series is usually the most difficult step of this procedure (see [6.11]).
• **Integral transforms**
The inversion procedure is based on classical integral transforms.
The Lynden–Bell ([6.22]) method is based on Laplace transforms, the Hunter ([6.16]) method on Stieltjes transforms, and the Dejonghe ([6.10]) method on Laplace–Mellin transforms. All these tools require the extension to the complex plane of the function $\rho = \rho(R, \Psi_T)$. Therefore, they are not meant to be applied to observational data. For a complete account of integral transform based methods, the interested reader is referred to the review by Dejonghe ([6.10]).

• Hunter–Quian method

This method is based on the theorem of Residues for functions of complex variables. The requirements on the function $\rho(R, \Psi_T)$ extended to the complex plane are much weaker than those for the integral transforms. The basic reference for this method is the paper by Hunter & Quian ([6.17]).

Unfortunately, the number of models for which the density and the associated DFs are known is disappointingly small.

12.3 Testing the consistency of multi–component OM models

From what was argued in Section 12.2, it would be very interesting to determine criteria for the DF to be positive. For spherically symmetric, multi–component OM models a theorem is derived below.

If a system is described by the superposition of different density components ρ_k, then *each f_k must be non–negative over the phase–space*. The theorem presented below allows us to check whether the DF of a multi–component spherical system, for which the orbital anisotropy of each component is of the OM form, is indeed positive, *without* an explicit calculation of it. We will deal with the radially anisotropic OM case, leaving the tangentially anisotropic case as a useful exercise for the interested reader.

Theorem 12.3a [Necessary, strong, and weak conditions for OM systems]
A necessary condition (NC) for $f_k \geq 0$ (see eq. [12.2b]) is:

$$\frac{d\varrho_k}{d\Psi_k} \geq 0, \quad Q_0 \leq \Psi_k \leq \Psi_k(0). \tag{12.3a}$$

For infinite systems ($Q_0 = 0$) for which the NC is satisfied, a strong sufficient condition (SSC) for $f_k \geq 0$ is:

$$\frac{d}{d\Psi_k} \left[\frac{d\varrho_k}{d\Psi_k} \left(\frac{d\Psi_T}{d\Psi_k} \right)^{-1} \sqrt{\Psi_T} \right] \geq 0, \quad 0 \leq \Psi_k \leq \Psi_k(0); \tag{12.3b}$$

finally, a weak sufficient condition (WSC) for $f_k \geq 0$ is:

$$\frac{d}{d\Psi_k} \left[\frac{d\varrho_k}{d\Psi_k} \left(\frac{d\Psi_T}{d\Psi_k} \right)^{-1} \right] \geq 0, \quad 0 \leq \Psi_k \leq \Psi_k(0). \tag{12.3c}$$

PROOF   We start by proving the NC (eq. [12.3a]). From eq. [10.7c] $d\varrho_k/d\Psi_T = 4\pi \int_{Q_0}^{\Psi_T} h_k(Q_k)dQ_k/\sqrt{2(\Psi_T - Q_k)}$, and if $h_k \geq 0$, eq. (12.3a) is obtained, because

$d\varrho_k/d\Psi_T = (d\varrho_k/d\Psi_k)(d\Psi_T/d\Psi_k)^{-1}$. The last factor is necessarily positive because the potential in spherical systems is monotonic[12.1]. The proof of the WSC (eq. [12.3c]) is trivial, and is obtained directly from the second identity in eq. (12.2b) for infinite systems. Again, the total potential is taken to be a function of Ψ_k. The proof of the SSC (eq. [12.3b]) is slightly more complex. For simplicity of notation let $I_g(Q) = \int_0^Q g(x)(Q - x)^{-1/2}dx$, with $x = \Psi_T$ and $g(x) = d\varrho_k/d\Psi_T$; from the first identity in eq. (12.2b), we look for a sufficient condition on the positivity of $I_g(Q)' = dI_g(Q)/dQ$. After changing the integration variable to x/Q, and after differentiation of $I_g(Q)$ with respect to Q, the original variable is restored, so that we get $I_g(Q)' = 0(1/Q)\int_0^Q [g(x)/2 + xg(x)'](Q - x)^{-1/2}dx$. Then a sufficient condition for $I_g(Q)' \geq 0$ is given by the positivity of the integrand, which can be rewritten as $\sqrt{(x)}d[g(x)\sqrt{x}]/dx \geq 0$. Setting $x = \Psi_T(\Psi_k)$ completes the proof.   ◁

This theorem can be restated using the radius instead of the potential as follows:

Theorem 12.3b [Necessary, strong, and weak conditions for OM systems]
The NC, SSC and WSC are given respectively by:

$$\frac{d\varrho_k(r)}{dr} \leq 0, \quad 0 \leq r \leq r_t, \tag{12.4a}$$

$$\frac{d}{dr}\left[\frac{d\varrho_k(r)}{dr}\frac{r^2\sqrt{\Psi_T(r)}}{M_T(r)}\right] \geq 0, \quad 0 \leq r \leq \infty, \tag{12.4b}$$

$$\frac{d}{dr}\left[\frac{d\varrho_k(r)}{dr}\frac{r^2}{M_T(r)}\right] \geq 0, \quad 0 \leq r \leq \infty. \tag{12.4c}$$

PROOF   The proof is obtained by changing variable from Ψ_k to r.   ◁

Note that the WSC is better suited than the SSC for analytical investigations, because due to the absence of the weighting square root of the total potential does not appear.

Some remarks are in order. The first is that the violation of the NC (eqs. [12.3a] and [12.4a]) is related only to the radial behavior of ρ_k and the value of r_{ak} (see eq. [12.2c]), and so this condition applies independently of whether any other component is added to the model. This condition is

---

[12.1] The formal expression for $\Psi_T(\Psi_k)$ is obtained by elimination of the radius between $\Psi_T(r)$ and $\Psi_k(r)$.

only necessary, thus f_k can be negative (and so the k–th component may be inconsistent) even for values of model parameters allowed by the NC. The reason is that the radial behavior of the integrand in eq. (12.2a) depends not only on the particular ρ_k and r_{ak}, but also on the total potential. Thus, a model failing the NC is *certainly* inconsistent, a model satisfying the NC *may be* consistent. Similarly, a model satisfying the WSC (or the more restrictive SSC) is *certainly* consistent, a model failing the WSC (SSC) *may be* consistent, because the conditions given by eqs. (12.4bc) and (12.5bc) are sufficient. As a result, the consistency of a model satisfying the NC and failing the WSC (or the SSC) can be proved only by direct inspection of its DF. For example, if one finds that for $r_{ak} \leq r_{ak}^{NC}$ the model is inconsistent (i.e., eqs. [12.3a] and [12.4a] are not verified), while for $r_{ak} \geq r_{ak}^{WSC} \geq r_{ak}^{SSC}$ the model is consistent (i.e., eqs. [12.3bc] and [12.4bc] are verified), this means that the true critical anisotropy radius (r_{akc}) must satisfy the relation $r_{ak}^{NC} \leq r_{akc} \leq r_{ak}^{WSC} \leq r_{ak}^{SSC}$. The second remark is that, in reality, the situation can be more complicated. In fact, for each component of the system, from eqs. (12.2ab) one can write

$$f(Q_k) = f_i(Q_k) + \frac{f_a(Q_k)}{r_{ak}^2}. \qquad (12.5a)$$

Let be A_+ the set defined by the property that $f_i > 0 \; \forall Q_k \in A_+$, i.e., the set of values of Q_k so that the isotropic component of the DF is positive definite. Then,

$$r_{ak} \geq r_{ak}^- := \sqrt{ \max\left\{ 0, \sup\left[-\frac{f_a(Q_k)}{f_i(Q_k)} \right]_{Q_k \in A_+} \right\} }, \qquad (12.5b)$$

is the condition to be satisfied in order to have a positive definite DF over A_+. Obviously, when $f_i > 0$ over all the phase–space (e.g., when the isotropic model is consistent) A_+ coincides with the total range of variation for Q_k, and $r_{akc} = r_{ak}^-$. In this case eq. (12.5b) shows that there is *at most* a lower bound for the anisotropy radius. This is the common situation encountered in one–component systems. When the set A_- (complementary to A_+) is not empty, i.e., $f_i < 0$ over some region of phase–space, a second inequality, derived from eq. (12.5a), must be verified:

$$r_{ak} \leq r_{ak}^+ := \sqrt{ \inf\left[\frac{f_a(Q_k)}{|f_i(Q_k)|} \right]_{Q_k \in A_-} }. \qquad (12.5c)$$

A general consequence of eqs. (12.5bc), applicable to *all* single or multi–component spherically symmetric, radially anisotropic OM models, is that the allowed region for consistency in the anisotropy space is given by $r_{\mathrm{ak}}^- < r_{\mathrm{ak}} < r_{\mathrm{ak}}^+$. Finally, note that if A_- is not empty and $f_a < 0 \; \forall Q_k \in A_-$, or $r_{\mathrm{ak}}^+ < r_{\mathrm{ak}}^-$, then the proposed model is inconsistent.

12.4 A fully analytical case

In this Section we will apply the derived NC, SSC, and WSC to the family of two–component (γ_1, γ_2) density–potential pairs. Each density distribution of the (γ_1, γ_2) models[12.2] belongs to the family of γ models (eq. [11.9j]):

$$\rho(r) = \frac{3 - \gamma}{4\pi} \frac{M \, r_{\mathrm{c}}}{r^\gamma (r_{\mathrm{c}} + r)^{4-\gamma}}, \qquad M(r) = M \times \left(\frac{r}{r_{\mathrm{c}} + r}\right)^{3-\gamma}, \qquad (12.6a)$$

where $0 \leq \gamma < 3$, M is the total mass and r_{c} a characteristic scale–length. The corresponding relative potential is given by

$$\Psi(r) = \frac{GM}{r_{\mathrm{c}}(2 - \gamma)} \left[1 - \left(\frac{r}{r + r_{\mathrm{c}}}\right)^{2-\gamma}\right], \qquad \Psi(r) = \frac{GM}{r_{\mathrm{c}}} \ln \frac{r + r_{\mathrm{c}}}{r}, \qquad (12.6b)$$

where the second expression holds for $\gamma = 2$. Note how for $r_{\mathrm{c}} \to 0$ the potential of the γ models is that of a central point mass (e.g., a black–hole) of mass M. In the following, the mass $M = M_{\gamma 1}$ and the characteristic scale–length $r_{\mathrm{c}} = r_{\mathrm{c}1}$ of the γ_1 model will be adopted as normalization constants, so that from eq. (12.6a) it follows that $\rho_{\gamma 1}(r) = \rho_{\mathrm{n}} \tilde{\rho}_{\gamma 1}(s)$ and $\rho_{\gamma 2}(r) = \rho_{\mathrm{n}} \mu \tilde{\rho}_{\gamma 2}(s, \beta)$, where $s = r/r_{\mathrm{c}1}$, $\rho_{\mathrm{n}} = M_{\gamma 1}/r_{\mathrm{c}1}^3$, $\mu = M_{\gamma 2}/M_{\gamma 1}$, and $\beta = r_{\mathrm{c}2}/r_{\mathrm{c}1}$. The key ingredient needed to recover the DF is the total potential $\Psi_{\mathrm{T}} = \Psi_{\gamma 1} + \Psi_{\gamma 2}$, where from eq. (12.6b) $\Psi_{\gamma 1}(r) = \Psi_{\mathrm{n}} \tilde{\Psi}_{\gamma 1}(s)$ and $\Psi_{\gamma 2}(r) = \Psi_{\mathrm{n}} \mu \tilde{\Psi}_{\gamma 2}(s, \beta)$, and $\Psi_{\mathrm{n}} = GM_{\gamma 1}/r_{\mathrm{c}1}$. With this choice, the structure of the (γ_1, γ_2) models is determined by fixing the four independent parameters $(M_{\gamma 1}, r_{\mathrm{c}1}, \mu, \beta)$, with the condition $\mu \geq 0$ and $\beta \geq 0$. The γ_1 component will be referred to as the *reference* component. In the particular cases of (1,1) and (1,0) models, the reference component is the $\gamma = 1$ model; for these cases, in order to simplify the notation, we define $(M_{\gamma 1}, r_{\mathrm{c}1}, \rho_{\gamma 1}, \Psi_{\gamma 1}) = (M_1, r_1, \rho_1, \Psi_1)$.

[12.2] For two particular cases, i.e., the (1,1) and (1,0) models, the DF and the solutions of the Jeans equations in the OM cases are obtained explicitly. The interested reader can find all the details in the two papers [6.4], [6.5].

Before discussing the complicated case of (γ_1, γ_2) models, we focus first on the general case of one–component γ models. In the following the unit mass and unit length are the total mass and the core radius of the model, with $s_a = r_a/r_c$.

Theorem 12.4a [NC for the γ models, OM systems] *For $2 \leq \gamma < 3$ the NC is satisfied for $s_a \geq 0$. For $0 \leq \gamma < 2$ the NC is satisfied for*

$$s_a \geq s_M \sqrt{\frac{2 - \gamma - 2s_M}{\gamma + 4s_M}}, \tag{12.7a}$$

where

$$s_M(\gamma) = \frac{4 - 5\gamma + \sqrt{(4 - \gamma)(4 + 7\gamma)}}{16}. \tag{12.7b}$$

PROOF   Through eq. (12.4a) the NC imposes a limitation on the anisotropy radius: $s_a^2 \geq s^2(2 - \gamma - 2s)/(\gamma + 4s)$, with $s \geq 0$. This is realized for s_a^2 larger than or equal to the maximum of the function on the r.h.s. of the previous expression. For $2 \leq \gamma < 3$ this is strictly negative, and so all values of s_a satisfy the NC. When $\gamma < 2$ the maximum is reached at the value given by eq. (12.7b).   ◁

Theorem 12.4b [WSC for the γ models, OM systems] *For $0 \leq \gamma < 3$ the WSC is satisfied for*

$$s_a \geq s_M^{3/2} \sqrt{\frac{3 - \gamma - s_M}{6s_M^2 + 2(1 + \gamma)s_M + \gamma}}, \tag{12.8a}$$

where, for $0 \leq \gamma < (\sqrt{73} - 5)/8 \simeq 0.443$

$$s_M(\gamma) = \frac{1 - \gamma}{3} +$$

$$\frac{2\sqrt{4 - \gamma}}{3} \cos\left[\frac{1}{3} \arctan \frac{\sqrt{(15 - 4\gamma)(3 - \gamma)(3 - 5\gamma - 4\gamma^2)}}{11 + 11\gamma - 4\gamma^2}\right], \tag{12.8b}$$

for $\gamma = (\sqrt{73} - 5)/8$

$$s_M(\gamma) = \frac{\sqrt{73} + 3}{8} \simeq 1.443, \tag{12.8c}$$

and, for $(\sqrt{73} - 5)/8 < \gamma < 3$

$$s_M(\gamma) = \frac{1 - \gamma}{3} + \frac{2(4 - \gamma)}{3s_0^{1/3}} + \frac{s_0^{1/3}}{6}, \tag{12.8d}$$

with

$$s_0 = (4 - \gamma)[11 + 11\gamma - 4\gamma^2 + \sqrt{(15 - 4\gamma)(3 - \gamma)(4\gamma^2 + 5\gamma - 3)}]. \quad (12.8e)$$

PROOF *The WSC (eq. [12.4c]) applied to the γ models gives the following inequality: $s_a^2 \geq s^3(3 - \gamma - s)/[6s^2 + 2(1 + \gamma)s + \gamma]$, for $s \geq 0$. After differentiation, one is left with the discussion of a cubic equation. Its discriminant is negative for $0 \leq \gamma < (\sqrt{73} - 5)/8$ and positive for $(\sqrt{73} - 5)/8 < \gamma < 3$. In the first case two of the three real solutions are negative, and the maximum of the r.h.s. is reached at the value given by eq. (12.8b). In the second case, by discarding the two complex conjugate roots, the maximum is reached at the value given by eqs. (12.8de).* ◁

A stronger limitation on r_a is obtained using the SSC, but, unfortunately, this condition for a generic γ leads to a transcendental equation that cannot be solved explicitly. However, we have the following

Theorem 12.4c [SSC for the $\gamma = 0, 1, 2, 3$ models, OM systems] *The SSC for $\gamma = 0$ gives:*

$$s_a \geq s_M \sqrt{\frac{3(1 + 2s_M - s_M^2)}{14s_M^2 + 10s_M + 2}} \simeq 0.501, \quad (12.9a)$$

where $s_M(0) \simeq 1.3149$. The SSC for $\gamma = 1$ gives:

$$s_a \geq s_M^{3/2} \sqrt{\frac{3(3 - 2s_M)}{28s_M^2 + 17s_M + 4}} \simeq 0.250, \quad (12.9b)$$

where

$$s_M(1) = \frac{s_0^{1/3}}{168} + \frac{1987}{56s_0^{1/3}} - \frac{3}{56} \simeq 0.9116, \quad (12.9c)$$

and $s_0 = 681939 + 84\sqrt{35887965}$. The numerical application of the SSC to the $\gamma = 2$ model gives $s_a \gtrsim 0.047$. Finally, the case $\gamma = 3$ is trivial, and the SSC reduces to $s_a \geq 0$.

PROOF The SSC (eq. [12.4b]) applied to the $\gamma = 0$ model gives the following inequality: $s_a^2 \geq 3s^2(1 + 2s - s^2)/(14s^2 + 10s + 2)$, for $s \geq 0$. The maximum of the r.h.s. of the previous equation can be obtained explicitly by solving a fourth-degree equation. The numerical value of the only physically acceptable solution is $s_M(0) \simeq 1.3149$ The SSC applied to the

$\gamma = 1$ model gives the following inequality: $s_a^2 \geq 3s^3(3-2s)/(28s^2+17s+4)$, for $s \geq 0$. After differentiation, by discarding the two complex conjugate roots of the resulting cubic equation, eqs. (12.9bc) are obtained. $\triangleleft$

We move now on to prove some results related to the consistency of (γ_1, γ_2) models:

Theorem 12.5 [WSC for (γ_1, γ_2) OM systems]
I) In the case of globally isotropic two–component $\gamma_1 + \gamma_2$ models with $1 \leq \gamma_1 \leq 3$ and $0 \leq \gamma_2 \leq \gamma_1$, the DF of the γ_1 component is positive over all the phase space, for all values of $(\mu, \beta) = (M_{\gamma 2}/M_{\gamma 1}, r_{c2}/r_{c1})$.
II) In the case of anisotropic γ models with a dominant BH at their center, it is possible to discuss analytically the limitations on r_a as a function of γ. For $3 > \gamma \geq 1$:

$$s_a \geq s_M \sqrt{\frac{(3-\gamma)(\gamma-2) + 4(3-\gamma)s_M - 2s_M^2}{12s_M^2 + 8(\gamma-1)s_M + \gamma(\gamma-1)}}, \qquad (12.10a)$$

where

$$s_M(1) = 2, \quad s_M(2) = \frac{(54+6\sqrt{33})^{1/3}}{6} + \frac{2}{(54+6\sqrt{33})^{1/3}} \simeq 1.2, \quad s_M(3) = 0.$$
$$(12.10b)$$

PROOF   The proof of point I) is conceptually straightforward but algebraically very cumbersome. Under the assumptions of point I) above, WSC is verified for all choices of (μ, β). In fact, when $3 > \gamma_1 \geq 1$ and $\gamma_1 \geq \gamma_2 \geq 0$, the WSC reduces to the discussion of the positivity of a rational expression, the denominator of which is strictly positive $\forall (\gamma_1, \gamma_2)$ and $\forall (s, \beta)$; the numerator factorizes in a strictly positive function and in a transcendental expression. By defining $\gamma_1 := 1 + \epsilon_1$ (with $0 \leq \epsilon_1 < 2$) and $\gamma_2 := \gamma_1 - \epsilon_2$ (with $0 \leq \epsilon_2 \leq \gamma_1$), the transcendental factor is:
$2(s+\beta)^4(1+1/s)^{\gamma_1}[6s^2 + 2(1+\gamma_1)s + \gamma_1] + \mu(s+1)^3(1+\beta/s)^{\gamma_2} F(s, \beta, \gamma_1, \gamma_2)$, where the first term is strictly positive, and $F(s, \beta, \gamma_1, \gamma_2) := 12s^3 + [(20 - 4\epsilon_1 + 4\epsilon_2)\beta + 8\epsilon_1]s^2 + [(10 + 5\epsilon_1 - \epsilon_1^2 + 5\epsilon_2 + \epsilon_1\epsilon_2)\beta + \epsilon_1\gamma_1]s + \beta\gamma_1(2 + \epsilon_2)$. In the range of values for ϵ_1 and ϵ_2 F is easily proved to be positive $\forall (s, \beta) \geq 0$. In point II), the assumption of a dominant BH is expressed algebraically by imposing in the WSC $M_T = M_{BH}$. After calculating the derivatives, we have to investigate for $1 \leq \gamma < 3$ the inequality: $s_a^2 \geq s^2[(3-\gamma)(\gamma-2) + 4(3-\gamma)s - 2s^2]/[12s^2 + 8(\gamma-1)s + \gamma(\gamma-1)]$, for $s \geq 0$. Note that for $3 > \gamma \geq 1$ the denominator is nowhere negative. After differentiation, one is left with the discussion of a quartic equation, and it can be shown that there exists only one maximum, located at $s_M(\gamma) \geq 0$. The explicit expression for $s_M(\gamma)$ is not very useful,

and so it is not recorded here. The explicit values of $s_M(\gamma)$ for $\gamma = (1, 2, 3)$ are given by eq. (12.10b).   ◁

Remark 13.2a

One particular case of point I) is the statement that globally isotropic (1,1) models can be consistently assembled for any choice of (μ, β). From point I) it also follows that the same is true for isotropic (2,2) models. Finally, if we consider that, for $r_c \to 0$, the potential of γ models becomes that of a point mass, the previous result means that a black–hole of any mass can be added at the center of a globally isotropic γ model with $3 > \gamma \geq 1$.

BIBLIOGRAPHY

1. Analytical Mechanics

[1.1] Arnold, V.I.: *Metodi matematici della meccanica classica*, (Editori Riuniti, Edizioni MIR), 1979

[1.2] Arnold, V.I. (ed.): *Dynamical systems. III*, (Springer-Verlag), 1988

[1.3] Fasano, A. & Marni, S.: *Meccanica analitica*, (Boringhieri), 1994

[1.4] Khinchin, A.I.: *Mathematical foundations of statistical mechanics*, (Dover), 1949

[1.5] Landau, L.D., & Lifsits, E.M.: *Meccanica*, (Editori Riuniti, Edizioni MIR), 1982

[1.6] McCauley, J.L.: *Classical mechanics. Transformations, flows, integrable and chaotic dynamics*, (Cambridge University Press), 1997

[1.7] Whittaker, E.T.: *A treatise on the analytical dynamics of particles and rigid bodies*, (Cambridge University Press), 1988

2. Celestial Mechanics

[2.1] Boccaletti, D. & Pucacco, G.: *Theory of orbits. 1: Integrable systems and non-perturbative methods*, (Springer), 1996

[2.2] Danby, J.M.A.: *Fundamentals of celestial mechanics*, (The Macmillan Company), 1962

[2.3] Meyer, K.R. & Hall, G.R.: *Introduction to Hamiltonian dynamical systems and the N-body problem*, (Springer), 1992

[2.4] Szebehely, V.: *Theory of orbits. The restricted problem of three bodies*, (Academic Press), 1967

[2.5] Wintner, A.: *The analytical foundations of celestial mechanics*, (Princeton University Press), 1947

3. Continuum Mechanics

[3.1] Aris, R.: *Vectors, tensors, and the basic equations of fluid mechanics*, (Dover), 1989

[3.2] Meyer, R.A.: *Introduction to the mathematical fluid dynamics*, (Dover), 1982

[3.3] Narasimhan, M.N.L.: *Principles of continuum mechanics*, (John Wiley & Sons), 1993

[3.4] Prigogine, I., & Nicolis, G.: *Self-organization in nonequilibrium systems*, (John Wiley & Sons), 1977

[3.5] Tassoul, J.-L.: *Theory of rotating stars*, (Princeton University Press), 1978

[3.6] Truesdell, C.A.: *Rational thermodynamics*, (Springer), 1984

[3.7] Truesdell, C.A.: *A first course in rational continuum mechanics. I*, (Academic Press), 1987

4. Potential Theory and Mathematical Methods

[4.1] Arnold, V.I.: *Equazioni differenziali ordinarie*, (Edizioni MIR), 1979

[4.2] Bender, C.M., & Orszag, S.A.: *Advanced mathematical methods for scientists and engineers*, (McGraw-Hill), 1978

[4.3] Bleinstein, N., & Handelsman, R.A.: *Asymptotic expansions of integrals*, (Dover), 1975

[4.4] Chandrasekhar, S.: *Ellipsoidal figures of equilibrium*, (Dover), 1987

[4.5] Chandrasekhar, S.: *An introduction to the study of stellar structure*, (Dover), 1967

[4.6] Fritz, J.: *Partial differential equations*, (Springer), 1991

[4.7] Gradshteyn, I.S., & Ryzhik, I.M.: *Tables of integrals series and products*, (Academic Press), 1965

[4.8] Jackson, J.D.: *Classical electrodynamics*, (John Wiley & Sons), 1975

[4.9] Kellogg, O.D.: *Foundations of potential theory*, (Dover), 1953

5. Stellar Dynamics

[5.1] Bertin, G.: *Dynamics of galaxies*, (Cambridge University Press), 2000

[5.2] Bertin, G., & Lin, C.C.: *Spiral structure in galaxies: a density wave theory*, (MIT University Press), 1996

[5.3] Binney, J., & Tremaine, S.: *Galactic dynamics*, (Princeton University Press), 1987

[5.4] Binney, J., & Merrifield, M.: *Galactic astronomy*, (Princeton University Press), 1998

[5.5] Chandrasekhar, S.: *Principles of stellar dynamics*, (Dover), 1942

[5.6] Ogorodnikov, K.F.: *Dynamics of stellar systems*, (Pergamon Press), 1965

[5.7] Saslaw, W.C.: *Gravitational physics of stellar and galactic systems*, (Cambridge University Press), 1987

[5.8] Spitzer, L.: *Dynamical evolution of globular clusters*, (Princeton University Press), 1987

6. Research and Review Articles

[6.1] Bertin, G., & Stiavelli, M. 1993, *Rep. Prog. Phys.*, **56**, 493

[6.2] Bertin, G., & Stiavelli, M. 1984, *A&A*, **137**, 26

[6.3] Binney, J. 1982 *ARAA*, **20**, 399

[6.4] Ciotti, L. 1996, *ApJ*, **471**, 68

[6.5] Ciotti, L. 1999, *ApJ*, **520**, 574

[6.6] Ciotti, L., & Bertin, G. 1999 *A&A*, **352**, 447

[6.7] Cuddeford, P., & Louis, P. 1995, *MNRAS*, **275**, 1017

[6.8] de Vaucouleurs, G. 1948, *Ann. d'Astrophys.*, **11**, 247

[6.9] de Zeeuw, T., & Franx, M. 1991, *ARAA*, **29**, 239

[6.10] Dejonghe, H. 1986, *Physics Reports*, **133**, 217

[6.11] Fricke, W. 1952, *Astron. Nachr.*, **280**, 193

[6.12] Gidas, B., Ni, W.-M., & Nirenberg, L. 1981, *Mathematical analysis and applications, Part A. Advances in mathematics supplementary studies*, **7A**, 369

[6.13] Hénon, M. 1959, *Ann. d'Astrophys.*, **22**, 126

[6.14] Hernquist, L. 1990, *ApJ*, **356**, 359

[6.15] Hubble, E. 1930, *ApJ*, **71**, 231

[6.16] Hunter, C. 1975, *AJ*, **80**, 783

[6.17] Hunter, C., & Quian, E. 1993, *MNRAS*, **262**, 401

[6.18] Jaffe, W. 1983, *MNRAS*, **202**, 995

[6.19] King, I.R. 1962, *AJ*, **67**, 471

[6.20] King, I.R. 1966, *AJ*, **71**, 64

[6.21] King, I.R. 1972, *ApJL*, **174**, 123

[6.22] Lynden-Bell, D. 1962, *MNRAS*, **123**, 447

[6.23] Merritt, D. 1985, *AJ*, **90**, 1027

[6.24] Michie, R.W. 1963, *MNRAS*, **125**, 127

[6.25] Miyamoto, M., & Nagai, R. 1975, *PASJ*, **27**, 533

[6.26] Ossipkov, L. P. 1979, *Pis'ma Astr. Zh.*, **5**, 77

[6.27] Plummer, H.C. 1911, *MNRAS*, **71**, 460

[6.28] Reynolds, J.H. 1913, *MNRAS*, **74**, 132

[6.29] Satoh, C. 1980, *PASJ*, **32**, 41

[6.30] Schwarzschild, M. 1954, *AJ*, **59**, 273

[6.31] Sersic, J.L. 1968, *Atlas de galaxias australes*, Observatorio Astronomico, Cordoba

[6.32] Wilson, C.P. 1975, *AJ*, **80**, 175

INDEX

"CompoMat" Loc. Braccone, 02040 Configni (RI), Italy
Finito di stampare dalla Nuova Grafica 86, nel mese di ottobre 2001